Industrial Maintenance
Reference Guide

The McGraw-Hill
Engineering Reference Guide Series

This series makes available to professionals and students a wide variety of engineering information and data available in McGraw-Hill's library of highly acclaimed books and publications. The books in the series are drawn directly from this vast resource of titles. Each one is either a condensation of a single title or a collection of sections culled from several titles. The Project Editors responsible for the books in the series are highly respected professionals in the engineering areas covered. Each Editor selected only the most relevant and current information available in the McGraw-Hill library, adding further details and commentary where necessary.

Industrial Maintenance Reference Guide

Robert C. Rosaler, P.E. Editor in Chief

James O. Rice Associates
New York

James O. Rice Associate Editor

James O. Rice Associates
New York

Tyler G. Hicks, P.E. Project Editor

McGraw-Hill Book Company

New York St. Louis San Francisco Auckland Bogotá
Hamburg London Madrid Mexico Milan
Montreal New Delhi Panama Paris São Paulo
Singapore Sydney Tokyo Toronto

Library of Congress Cataloging-in-Publication Data

Industrial maintenance reference guide.

(McGraw-Hill engineering reference guide series)
"The material in this volume has been published previously in
Standard handbook of plant engineering, by Robert C. Rosaler and
James O. Rice"—Verso of t.p.
Includes index.
1. Plant maintenance—Handbooks, manuals, etc. 2. Industrial
equipment—Maintenance and repair—Handbooks, manuals, etc.
I. Rosaler, Robert C. II. Rice, James O'Neill, 1910–
III. Hicks, Tyler Gregory, 1921– . IV. Standard handbook of
plant engineering. V. Title VI. Series.
TS192.I52 1987 658.2′02 87-3167

ISBN 0-07-052162-X

1234567890 DOCDOC 89210987

ISBN 0-07-052162-X

Printed and bound by R. R. Donnelley and Sons

Contents

Preface, vii
Contributors, ix

Preface

This is a concise guide to equipment maintenance, repair, lubrication, and corrosion prevention. The guide will be useful to plant engineers, mechanical engineers, electrical engineers, plant operators, plant superintendents, cost estimators, schedulers, factory personnel, drafters, and engineering students. A condensation of Rosaler and Rice—STANDARD HANDBOOK OF PLANT ENGINEERING, the guide has numerous illustrations to guide the reader to the correct procedure for a given situation in maintenance.

Key topics covered include: the maintenance facility; selection, use, and care of basic tools; welding, cutting, brazing, and soldering; metal resurfacing by thermal spraying; structural adhesives; diagnostic instrumentation for machinery maintenance; lubricants and lubrication systems; causes and control of deterioration; paints and protective coatings.

Each sectional topic was prepared by one or more authorities in the field. Hence, the guide presents valuable data and methods that can be used with confidence.

To expand the information presented, most sections contain a bibliography and listings of sources of additional data. Readers will find these especially helpful when they want to do more research on a given topic.

Each section is presented in a way that permits easy use. For example, the section on Solid Lubricants contains these subheadings, in this order: Introduction; Uses; Forms and Applications; Characteristics; Summary; Bibliography. Both USCS and SI units are used throughout the guide.

This guide will be a helpful daily companion to anyone performing maintenance work of any kind. It also will be of much help to maintenance managers in all types of plant activities.

Where applicable, the importance of energy conservation and pollution control are emphasized in the topics discussed. With such an orientation, the content of this guide meets the demands of industry today.

Tyler G. Hicks, P.E.

Contributors

Anne Bernhardt
Staff Engineer
Gulf Research & Development Co.

Howard Cary
Vice President, Welding Systems
Hobart Brothers Co.

Marvin Campen
Product Manager, Synthetic Fluids
Gulf Oil Chemicals Company

Rodger C. Dishington
Sr. Engineer
Imperial Oil & Grease Company

Richard J. DuMola
Materials Engineer
METCO, Inc.

J. Mark Gilstrap
Technical Manager, Mechanical
  Engineering Services
Bently Nevada Corp.

William S. Harrison, P.E.
Director of Technical Services
Engelhard Industries Division
Engelhard Corp.

Walter D. Janssens
Director, Solid Research
Imperial Oil & Grease Company

Robert J. Klepser
Sr. Development Chemist
Heavy Duty Maintenance Coatings
PPG Industries, Inc.

Philip H. Maslow, P.E., FCSI
Consultant
Chemical Materials for Construction

William M. Scardino, P.E.
Consulting Engineer

Jack P. Waite
Manager, Engineering Field Services
Imperial Oil & Grease Company

section 1

Equipment for Maintenance and Repair

chapter 1-1

The Maintenance Facility

by
William S. Harrison, P.E.
Director of Technical Services
Engelhard Industries Division
Engelhard Corp.
Plainville, Massachusetts

INTRODUCTION

Every maintenance department needs a well-planned workplace. Though large machine repairs are usually done on site, many components may need to be returned to the shop for repair. When possible, removal of equipment from the production floor is more likely to yield a better, more complete repair or rebuild. The nonrepetitive nature of the tasks demands a location where skill and tools can be combined and applied without the distractions of production.

Plan your facility to suit the size and type of responsibilities of your firm. The shop in a textile mill with looms will be equipped differently than that in a chemical plant. Common factors to consider: basic work area; workbenches; mobile or permanent toolbox area; parts-cleaning facility; basic machine tools; access for hoists, cranes, or forktrucks; spare parts and raw-material storage; welding area; electrical area; ventilation-fume exhaust; utilities—air, gas, water, etc.

Specialized areas such as instrument repair, pipe fitting, plumbing, blacksmithing, and others are added to suit. It is seldom advisable to combine the functions of model shop, development shop, or gauge control in the maintenance shop because of the difference in work content.

There are many workable layout variations. Access, manipulation in disassembly, and temporary storage of components are important. Some operations may require ladders, stands, or scaffolding. Vehicle or forktruck repair requires pits or lifts and special ventilation.

If space allows, the following ideas should be considered:

1. An area separated from production is preferable.
2. A main aisle for access with workplaces on each side is efficient.
3. Benches can be used to delineate the individual stations at right angles to the aisle or placed parallel to the aisle at a distance equal to the depth of the workplace.
4. Electrical and cleaning stations should be out of the way or at the perimeter.
5. Parts need racks or bins; fencing may be needed for security; some items can be double-decked for space saving.
6. Pipe and steel racks should be accessible to handling equipment.
7. Offices should be accessible, but out of the way—at the perimeter, on the mezzanine, etc.

Think ahead! Items may come in piecemeal but may have to go out assembled. When planning the facility in a very large plant, such as automotive, aircraft, or appliance, employ the *satellite concept*. using small shops equipped to handle the more routine demands of a particular area.

BASIC TOOLS

Basic tools are common hand tools, powered or manual and usually portable. The simple drill press, grinder, and small lathe are generally included as are common electrical tools like a voltage tester, multimeter, megger, etc.

Selection criteria for tools include:

1. **Cost**   Always important.
2. **Quality**   Remembered longer than cost.
3. **Design**   Feel, ease of use, functionality.
4. **Size**   Be conservative, do not overload any tool or instrument.
5. **Reliability-Durability**   Works consistently and can stand hard use.
6. **Reputation of the maker**   Can be meaningful.
7. **Repair facilities**   Proximity; time is valuable.

In general, there are some quality and capability differences between hand tools used at home and those in the factory. However, power and machine tools are substantially different. It seldom pays to mix the two. The timely success of a department should not depend on weekend- or home-type tools.

LARGER FACILITIES

Beyond the basic hand tools necessary in every shop, the plant engineer must consider work demand and cost. The cost and complexity of use of some modern tools may dictate that you buy the service instead. Your decisions will depend on the following criteria:

1. Cost of the tool
2. Frequency of use
3. Space requirements
4. Skill requirements
5. Priority of need

The last is difficult to assess, given the demanding nature of the profession. Avoid the tendency to collect unique tools that are rarely used. Only the larger and/or specialized shops can justify diamond-sawing and core-drilling equipment for concrete work or the latest infrared scanning rigs for predictive plant-maintenance work.

After the basic considerations come the functional requirements of each plant which dictate specialized tool needs. A detailed list for each industry is beyond the scope of this book. Apply the basic criteria given above to each facility. Implications to be considered include:

1. Age of plant machinery
2. Reaction time expected by management
3. Value of lost production time
4. Availability of outside services
5. Training required to operate sophisticated equipment

Management may lean toward minimizing an investment in costly, specialized tools since it has a negative effect on return on invested capital, particularly if the service can be bought. Plan acquisitions that take into account prevailing ideas.

Other acquisitions to be considered include magnetic-base drill press, wire-feed welding equipment, cleaning stations with pumped solvent, steam-cleaning rigs, sand-blasting gear, drill press or radial drill 1 in (25 mm) up, Bridgeport ®* mill or K & T Universal, surface grinder, flame-cutting table and gear, impact drills to 1½- or 2-in (38- or 51-mm), pneumatic chisels (street busters), optical alignment equipment, electronic balancing equipment, vibration analysis equipment, portable breathing equipment, gear pulling and/or portapower rigs, industrial vacuum cleaners, ground detection testers, thermocouple calibration galvanometers, portable generators, portable hydraulic personnel lifts, sound measurement frequency analyzers, environmental gas-liquid test equipment, and power hack and band saws.

LIGHTING AND POWER†

Whether planning a new facility, revamping an old, or just relocating within a plant, the steps are the same. Adopt a methodical approach to avoid avoid costly errors. Electrical work can represent a sizable fraction of total project cost, so:

1. Analyze the power requirements for existing and projected equipment.
2. Allow for expansion.
3. Allow for temporary power to equipment being repaired.
4. Decide on a lighting requirement.
5. Carefully integrate the light and power design with the plant layout.

Medium to larger maintenance facilities in plants fewer than 30 years old should have, and most probably will have, 460-V-ac, three-phase power available. This is the *standard* industrial *low*-voltage distribution. A four-wire, Y configuration is good since wiring economy can be effected with 277-V fluorescents if they are the lighting choice.

Even the most modern plants need 230-V, three-phase and 115/230-V, single-phase power. These are easily obtained via single or three-phase stepdown transformers. This is generally termed *miscellaneous power* and is used for smaller machine tools or for power equipment being repaired, tested, or constructed.

There are no firm rules for sizing the main electric feeder to a maintenance facility. The task requires professional expertise. As a rule of thumb a *demand factor* (ratio of maximum demand to total connected load) for the average facility is probably in the

---

*Registered trademark, Bridgeport Machine Works.

†See also Sec. 3 of Rosaler and Rice: *Standard Handbook of Plant Engineering*, 1983.

range of 0.2 to 0.3. It is important to consider again whether most of the work is within the facility or elsewhere.

Only extremely specialized shops would require electric service at higher voltages. Such installations should be left to specialists in the profession.

A typical installation will have:

1. Machine and equipment loads
2. 117-V single-phase receptacles
3. 208- or 230-V, single-phase receptacles
4. 208- or 230-V, three-phase receptacles
5. 460-V, three-phase receptacles
6. Special receptacles for welders, etc.
7. Lighting circuitry and switching

In a new plant, try to include special receptacles for welders and other equipment at multiple locations *in the work areas* to save time and money.

How do you electrify a facility? The scheme depends on the size, quantity, and location of the equipment. Machines can be fed from power distribution panels, busways, miscellaneous power panels, motor control centers, connection boxes, and multiple fused disconnects. When recommended practice is followed, all conform with the **National Electrical Code®**.

Provide more than adequate light in the work area; 100 footcandles (fc) at bench level is a good rule. This may be supplemented by fixtures installed directly over workbenches with local control. In a new facility, high-intensity sodium vapor fixtures are most effi-

*National Electrical Code® is the trademark of the National Fire Prevention Association, Quincy, Mass.

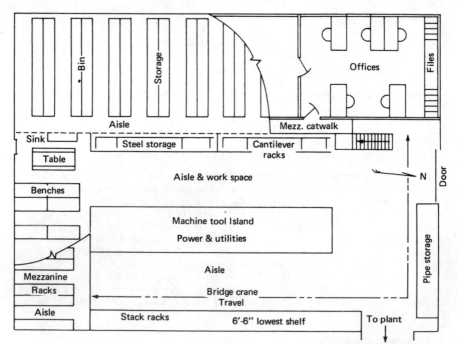

Figure 1-1 A compact contralized maintenance facility, 58 × 78 ft (17.5 × 24 m).

cient if color is not a factor. The time-proven fluorescent is available in a design using 25 percent less than normal energy for comparable lumen output.

Provide portable stands with a bank of floodlights to use in the facility or out on the job. If the normal mode of operation is not centered in the facility, consider switching. This can yield overall levels of illumination only as required. Efficient use of energy is a must.

A FACILITY IDEA

There are many ways of fitting together the necessary elements of an efficient facility. Figure 1-1 shows the layout of a shop serving a metalworking factory of over 250,000 ft^2 (23,200 m^2) and serviced by 30 to 35 technicians

Several features in the figure are:

1. One side is occupied by 7-ft (2.13-m)-high bins with a controlled-access aisle connecting all bin aisles.
2. Above the bins and supported by them are supervisory and plant engineering offices.
3. On the end is a workbench, lunch, and washup area.
4. Above No. 3 is a mezzanine with rack storage for more bulky supplies.
5. The second side has stacking racks of 4-ft (1.22-m) depth with the first shelf at 6 ft 6 in (2 m) to allow tool-chest storage and work space for grinders and sanders. On the wall, not shown, is a V-belt rack. Storage continues to an eave height of 14 ft 6 in (4.4 m).
6. The remaining end has a long pipe rack with multiple levels and bulk storage on top. A door occupies the remaining length.
7. Two aisles, one 9 ft (2.7 m), the other 10 ft (3 m), flank the machinery *island*. The wider aisle allows a truck to drive in and be unloaded by a 2-ton (1820-kg) bridge crane which sweeps the entire working area. The aisles provide forktruck access and working area. Only portable units or equipment under repair are allowed in the aisles.
8. Machine tools are installed back-to-back with electric power at three voltages and single- and three-phase in between. The wireway run at head height does not interfere with the sweep of the bridge crane. Power panels, transformers, etc., are located on the south wall. An air line also runs between the machines.

An arrangement to be workable and efficient should consider storage and work space. It should have maximum density for storage with easy access and a maximum open work area with hoisting and handling service.

BIBLIOGRAPHY

Muther, Richard: *Systematic Layout Planning,* Cahners, Boston, 1976.
Dergamo, E. Paul: *Materials and Processes in Manufacturing,* Macmillan, New York, 1957.
Ball, John E.: *Carpenter's & Builder's Library,* vol. 4, Audel, Indianapolis, 1965.

chapter 1-2

Selection, Use, and Care of Basic Tools

by
William S. Harrison, P.E.
Director of Technical Services
Engelhard Industries Division
Engelhard Corp.
Plainville, Massachusetts

INTRODUCTION

In earliest days, the use of tools distinguished the intelligence and evolution of the human race. Tools multiplied productivity, and this is still the objective in the acquisition and use of good tools.

Craftspeople must have the proper tools to accomplish their work safely, quickly, and correctly. In most cases involving machinery maintenance, it is impossible to complete a task without a substantial number of common hand tools and several special pullers or shaped wrenches.

When individuals and companies depend on tools for their existence, it pays to select good, even the finest, tools. It takes very little experience to perceive the difference in *feel, balance, fit,* and *design for the job* of a fine tool or tool assembly.

Virtually all fine hand tools and many of their component parts are manufactured from alloy steel forgings. It is the continuous-grain skeleton following every curve and shape intricacy that provides the strength within a relatively light section. One has only to try to fit the oversize gripping hook of a cheap puller in a tight space behind a gear to appreciate the design and strength of the forged, alloy steel version of the same tool. Or contemplate that delicate hooked probe that a dentist inserts between your teeth. As a professional yourself, select the best you can find. Price is not the only criterion, but fine tools are seldom sold as loss leaders.

Even a small shop will need an extensive assortment of hand tools, portable electric tools, and a few basic machines. If you must live within a tight budget, be even more selective. The fine tool can usually be *stretched* over a wider range of application than the poorer grade.

The best source for tools is an industrial-supply or mill-supply distributor. Look for one with a good reputation for service and repair since all power tools will eventually require maintenance. Several complete tool lines are sold directly rather than through local distributors, for example, those sold by Snap-On Tool Corp. and Sears Roebuck and Co.

Our purpose is to present a guide or checklist for selection. Rather than use space for pictures of the many types and sizes, we recommend that each shop keep on hand the catalogs of the many reputable manufacturers. These can be acquired either through their distributors or directly. Use them for more detailed information on the tools listed and briefly described here.

SCREWDRIVERS

Types

The types of screwdrivers are:

1. **Standard**   Common slot, sized by blade width, ⅛ to about ¾ in (3 to 19 mm) (larger sizes less used as slots disappear in favor of socket heads).
2. **Phillips**   Cross-slotted, tapered from tip, sized 0 to 4.
3. **Reed Prince**   Variation of Phillips, shallow taper from tip, little or no vertical taper.
4. **Clutch Head**   Two intersecting circles when viewed head on, sized by the wider dimension.
5. **Auto-Return, Ratcheting**   Spiral-grooved shafts or gear-and-pawl mechanisms to enable semicontinuous turning without changing hand grip, sized by the blade. Many adapt to a range of interchangeable blades.
6. **Screw-Holding or Magnetic**   Spring steel fingers grip under screw head; two-piece blade is offset to bear tightly against both slot sides; entire blade is permanently magnetized.

7. **Precision or *Jewelers'*** Slot type usual, Phillips more rare, blade width in decimal inches like 0.040, 0.070 in (1 to 2 mm) for tiny screws like those found in watches, instruments. Generally with free rotating finger rest on top and knurled steel body.

Selection and Use

Fine screwdrivers have forged steel blades or ends. Blade ends are ground to the proper size. Choose the size that fits snugly with little clearance. Any tendency to rock in the slot will ultimately wear the blade or round off the corners. Never use a screwdriver for a wedge, prybar, or punch. Should a blade tip break, it is a tedious task to grind it to the correct angle and size without using a vise and surface grinder. It is preferable to replace a broken screwdriver since hardness and temper of the blade would likely be affected by regrinding.

WRENCHES

The term *wrench* is generic, and there are literally hundreds of tools that hold, grip, and turn. Whether needed for gripping nuts, cap screws, pipe, shafts, or other components, choose a tool of proper gripping capacity and leverage. Never extend capacity with pipe extensions over a handle. If the jaw or grip breaks, accidents are quite likely to occur when an individual is exerting an undesigned pull or push on the tool. Observe manufacturers' specifications as to capacity and safe torque limits. Overloading can distort a jaw, grip, or socket, making the tool useless and in many cases *peening, rounding,* or *damaging* the fastener or device being loosened.

Types

Most common wrenches are available in both U.S. customary and metric sizes:

Open End. $\frac{3}{16}$ to 2 in (4 to 50 mm)

Box End. $\frac{3}{16}$ to 2 in (4 to 50 mm)

Combination. Open and box combined

Open End. One end grip, smooth tapered handle

Socket. $\frac{3}{16}$ to 4 in (5 to 200 mm) with drive $\frac{1}{4}$ to $1\frac{1}{2}$
    Where access allows, choose a wrench that contacts all sides of a fastener or object. For a hex head, a 6 point is always preferable to 12 point.

Adjustable End. 4 to 24 in (100 mm to 0.6 m) length

Stillson. 8 in to 6 ft (0.2 to 1.8 m) length, adjustable, serrated jaws for pipe work

Monkey. 6 to 24 in (150 mm to 0.6 m) length, adjustable, smooth jaws, generally made obsolete by adjustable end wrenches

Box for Tubing Application. $\frac{3}{8}$ to $\frac{3}{4}''$ (9.5 to 19 mm); essentially a six-point box-end wrench with a section of the perimeter cut away to permit entry of a thin-wall tube and access to a tube fitting when it is connected. Prevents *rounding* of soft brass fittings.

HAMMERS

There are three basic types of hammers, and most of the many special types are adaptations of the basic types for the work requirement. The basic types are curved-claw (carpenter style), ball-peen (machinist style), and maul or sledge.
    In addition to steel, modern hammers use materials such as rubber, filled plastics, and other nonmetallics for the heads; steel, wood, or reinforced plastic is used for the handles. Hickory is still the prime handle material for its shock deadening and *balance,*

though fiberglass-reinforced plastic has similar qualities and does not readily chip from a missed blow.

Proper care begins with using a hammer of correct weight for the task; position yourself so that a direct, well-aimed blow of proper force can be struck. Handles should never be used for prying or as levers. Touch up minor nicks on the striking face and head with a file. Keep the handles clean to prevent slipping or turning in the hand. Sand chipped or splintered wood. Handles may be replaced, but this painstaking process is almost a lost art. If in doubt, do not take the chance of using a hammer with a loose head; this is an extremely dangerous condition. Always wear safety glasses.

Types

Curved-Claw (Carpenter's Hammer)

This type is made with heads varying in weight from 4 to 24 oz (113 to 680 g); 16 or 20 oz (450 or 570 g) is the preferred weight. Curved-claw hammers have a flat face for nail striking. Avoid extraction of nails beyond 16d; use a nail puller or wrecking bar which is better suited for the task.

Ball-Peen (Machinist's Hammer)

This type of hammer varies in weight from 4 to 40 oz (113 g to 1.1 kg) or more; 12 or 16 (340 or 450 g) is the preferred weight. Ball-peen hammers have a crowned face for striking and a peen end for upsetting pins, rivets, etc. Surfaces are rounded to minimize chipping and denting since these tools generally strike metals of hardness comparable to their own.

Maul or Sledge

The heads of mauls vary in weight of 4 to 20 lb (1.8 to 9.1 kg). Usually they are double-faced with a crown. They are generally used for striking or driving large keys or shafts and are used for masonry demolition only as a last resort.

Other Types

Special types include bricklayer's, mason's, upholsterer's (tack), shingle, cobbler's, special sheet-metal forming, and a variety of half-hammer, half-axe, or half-hatchet combinations.

DRILLS

Most drills used for maintenance are electric since hand drills are not practical in metalwork except as noted. Keyed Jacobs chucks are used on all but the lightest tools, and handle bits up to ¾ in (19 mm). With occasional cleaning and oiling, they are completely reliable. Tools should be kept clean and oiled as recommended and not overloaded. Stalling is a sign of overload and will cause burning of the commutator and wear on brushes. Industrial-quality drills usually have a simple system for brush replacement, and a yearly change is good preventive maintenance.

Common practice in the manufacture of industrial twist-drill bits is to supply a straight shank equivalent to the diameter of the bit up to ½ in (12.7 mm) diam. Beyond ½ in, a Morse taper shank is more common. The taper on all sizes from 0 to 7 is approximately ⅝ in/ft (15.9 mm per 305 mm). This taper develops so much frictional resistance that it is termed *self-holding*. The small end terminates in a flat-sided tang which fits in a matching female slot in the chuck or socket. This is used to remove the bit via a tapered key or wedge which is inserted in the open slot and tapped lightly, thus driving the bit out of the taper. Refer to a *machinists' handbook* to determine which size taper applies to a specific bit diameter.

While straight-shank twist drills *are* available in the larger sizes, another advantage of the Morse taper is its self-centering and positive alignment. It is advisable to specify drilling equipment, particularly drill presses, to include Morse taper chucks or sockets for diameters over ½ in (12.7 mm).

Hand Drills

The smallest hand drill is called a *pin vise* and has a straight knurled shank and tiny hand chuck to hold bits in *jewelers' sizes*. It is rotated by finger action. Geared hand drills are made with straight or pistol-grip handles with chucks of ⅛- or ¼-in (3.2- or 6.35-mm) capacity. Larger sizes are termed *breast drills* from the shaped plate which is borne on the chest while the crank handle is being turned. Most of these units are approaching antique status. A better bet is a *cordless* or battery-powered unit.

Electric Drills

All shops require several electric drills of different sizes and styles. Factors to consider in their selection are discussed in the following paragraphs.

Style
Pistol-grip drills are most common and can be used for most applications. Some heavier versions have a side handle to apply extra force; an end handle may be preferred in heavier, repetitive drilling; a spade handle is common from ½ in (12.7 mm) and up and accompanied by a companion side handle to allow two-hand control and force; extended spade and side handles are found on drills ¾ in (19 mm) and up to allow two or more individuals to hold and push the drill. For tight clearances there are right-angle styles that are particularly useful for electricians and plumbers.

Capacity
Capacity is defined as the largest-diameter drill bit that the machine will accept. Capacity ranges are usually ¼, ⅜, ½, ⅝, ¾, 1, 1¼ in (6.35, 9.5, 12.7, 15.9, 19, 25.4, 31.9 mm).

A practical minimum bit size is 1/16 in for the ¼-in chuck. Special drills and chucks are available which will handle numbered sizes to 97, but these would find little use in maintenance. Material to be drilled will govern speed and power requirements and must be considered with capacity.

Speed
It is important to know the material to be drilled and to match this using handbook formulas for the required speed within broad limits. In general, harder metals like nickel, stainless steel, and carbon steel require the lowest speeds. Softer metals such as aluminum, magnesium, and zinc require medium speeds, and most woods require high speeds. As an example, applying this to a drill of ¼-in (6.35-mm) capacity, speeds of about 1500, 3500, and 5000 r/min would represent reasonable compromises. The newer variable-speed or multirange tools may yield some flexibility, though it is very difficult to judge and control their speed.

Power and Construction
Industrial-quality drills are designed with sufficient power for the particular size. Close comparison will reveal some with slightly more power than others. Before choosing, examine the other features. Look for head-treated gears, ball or roller bearings, replaceable brushes, and a well-molded case. Expect these industrial-quality, heavy-duty portable drills to cost anywhere from 5 to 10 times what you might pay for a home-workshop type. They are built to last and are usually worth repairing when they do falter. Even the smallest shop would have use for two ¼-in (6.35-mm) drills, low and medium speeds, a ½-in (12.7-mm) low-speed unit, and a ¾-in (19-mm), 300- to 400-r/min unit.

Magnetic Drill Press

Medium to large shops should consider a magnetic-base portable drill press. These units make possible the drilling of clean, correctly aligned holes in locations where individuals could not hold and apply force to a hand drill. The geared feed handle yields point pressures comparable to those of a regular nonportable drill press.

Drill Stands

These are small, portable, column-and-stand, lever-feed units that hold portable drills. They are handy for repetitive drilling and provide better alignment where throat capacity allows their use.

Hammer Drills and Impact Drills

These drills combine rotational motion of the bit with repeated vibratory hammer blows. This type of drill is mandatory when drilling concrete, brick, or other refractory-like material and is used in combination with carbide-type drill bits. Do not use carbide bits with regular, nonimpact drills in concrete or similar materials. This is a quick way to ruin the drill or its gears and bearings.

ELECTRIC HAND TOOLS

Depending on the type and size of shop and the capability you are attempting to build you may consider some of the following electric portable hand tools:

Types

The types of electric hand tools are:

Screwdrivers-tappers	Grinders, disk-type
Impact wrenches (also air-driven)	Grinders, wheel-type
Routers, wood or metal	Grinders, die high-speed
Shears	Percussion hammers
Nibbers	Demolition hammers
Circular saws	Portable planers
Reciprocating hacksaws	Riveters
Saber saws or jigsaws	Chisels
Portable band saws	Nail guns
Sanders, belt-type	Staple guns
Sanders, disk or rotary	Electrohydraulic pull-and-squeeze units
Sanders, reciprocating	

Selection and Use

The same rules of selection apply here. Examine capacities and match them to needs. Many distributors will demonstrate tools or will even lend tools for trial. All these specialty tools are potentially dangerous. Follow the manufacturer's instructions, and proceed slowly and carefully. If possible, learn from a worker who is already skilled in the use of the tool. Many of the more powerful units develop substantial force reactions in operation or if they bind or jam. *Make certain you have the proper footing and stance to resist these forces. Wear safety glasses.*

Clean tools after use. Impact or percussion tools usually require special lubrication procedures.

The electrohydraulic pull-and-squeeze units are extremely versatile. They employ compact hydraulic pumps which supply fluid under high pressure to cylinders of various strokes. The cylinders combine with puller jaws and grips to remove press fit bearings, sleeves, and gears. They generally come in sets with assorted special tools to handle a wide range of applications where pulling or pushing forces ranging up to 10 tons (9 metric tons) are required. A diverse set may cost half a year's craft wage, but it is a worthwhile investment.

DRILL PRESS

Perhaps the most widely used machine tool in the shop, the drill press is a versatile unit that allows proper matching of speed and feed for the drill bit and the material being penetrated. The types applicable to plant engineering and maintenance are bench, floor, upright, and radial. The essential size or capacity of a drill press is described by the following parameters.

Drill Press Parameters

Size

Size is measured by the distance from the centerline of the spindle-chuck-quill to the outside or nearest point of the column. In effect, it is the diameter of the largest disk in which a center hole could be drilled.

Capacity

The capacity is the maximum diameter of the bit that can be inserted in or driven by the chuck or bit holder.

Travel

Travel is defined as the vertical distance that the bit can be fed via movement of the quill and spindle assembly.

Speeds

Individual speeds may be selected through pulley or gear combinations to match bit diameter and material hardness.

Feeds

Different vertical feed rates, in inches per revolution, are available via the power train.

Types of Drill Presses

Bench Type

This is generally limited to a size of 15 in (380 mm) and a travel of 4 in (102 mm). It is limited in application by the distance from the bit to the pedestal.

Floor Type

This is essentially the same as the bench type except for the longer column and greater clearance for a higher or larger workpiece. The smallest shop should include a floor-type press with capacity to ¾ in. (19 mm) via a Morse taper chuck. Versatility of this unit would be augmented if it included power feed and a minimum size of 18 in.

Upright

Generally upright implies a floor-type press with heavy box or tube column and power feed.

Radial

A radial drill is one consisting of a large-diameter round column fixed to a base; an arm is attached to the column and may rotate about it; the arm may be raised or lowered; a drilling head moves radially on the arm via a lead screw; the head provides multiple spindle rotation (speeds) and feeds. It is very versatile for drilling multiple holes in large, difficult-to-move workpieces. Size is specified via column diameter and radial travel of the head.

A radial drill press affords the capability to machine larger machined components whose dimensions exceed the capacity of regular drill presses.

GRINDER

With the exception of portable or off-hand grinders, the most important function of grinders in the maintenance shop is to sharpen the cutting tools of other machines.

Types

The types most used in our applications are:

Surface Grinder with Reciprocating Table and Horizontal Spindle

This is the workhorse of the shop in terms of capability. It can be used to sharpen many types of cutters with the different-shaped grinding wheels that are available. The grinding wheel rotates on a horizontal spindle as the table reciprocates or travels right to left (and back) in a horizontal direction. It is excellent for establishing the thickness of small, critical parts.

Surface Grinder with Rotating Table and Vertical Spindle

This is commonly called a Blanchard grinder. It is used for clearance purposes to establish parallelism between two critical surfaces or to establish a bearing or mounting surface. This type of surface grinding is a prime restoration step in many maintenance operations.

Universal Tool and Cutter Grinder

This is an ingenious adaptation of the common horizontal-shaft shop grinder that employs manual table reciprocation as well as vertical and rotational adjustment of the horizontal wheelhead to enable proper access to virtually all types of cutters with the ability to hold or grind proper rake, lip, nose, and relief angles. It is a versatile but somewhat specialized tool.

Portable Grinders

Among the often-used types is the angle-head, hand-held grinder-sander which accepts a variety of abrasives and has provision to attach appropriate wheel guards. Typical abrasives are a thin, flexible disk on a rubber wheel; a thicker, rigid disk; a formed, or dished, disk; a common cup shop wheel; and a wire brush. Portable grinders handle many operations such as weld grinding, descaling, paint preparation, etc.

Handy for certain finer work is the small, high-speed hand grinder which accepts mandrel-mounted abrasive *stones*. These are often called *die grinders*.

Selection

Grinding speed is quite critical. The machine must closely match the grit and size of the wheel or stone; otherwise the abrasive will not cut properly or will tend to *load* up. More important, there is a danger of wheel disintegration, particularly at improper low speeds. Safety glasses, preferably with side shields, are a *must* in any grinding or sanding operation. Gloves and long sleeves are essential with the portable angle grinders to protect from sparks, grit, dust, or loosened particles.

SANDER

Sanders are useful in many phases of building maintenance, carpentry, or cabinet work; they are more necessary in larger shops than in smaller ones.

Types

Disk and Belt Sanding Machine

This is a nonportable bench- or stand-mounted machine specifically designed for finish woodworking. It generally combines a 10- or 12-in (254- or 305-mm)-diam disk wheel with a 6- or 8-in (152- or 203-mm)-wide belt in combination with appropriate tables to hold and feed the work.

Portable Belt Sander

This is available with a 3- or 4-in (76- or 102-mm) belt width. It is ideal for rough sanding of wood with high removal rates. Some models include self-contained dust-collection systems.

Reciprocating-Oscillating Sander

This is available in several pad sizes (effective area) from about 2 x 5 to 4½ x 9 in (51 x 127 to 114 x 228 mm). Generally, it is most effective for fine or finished woodwork.

Use

Most important with sanders is the rule, "Let the paper do the work." Little, if any, additional pressure is necessary. Choose *open-coat* abrasives that do not load. Synthetic aluminum oxide abrasives have a longer life than flint or garnet abrasive.

LATHE

The lathe is the basic machine for metal turning and is important to even the smallest shop. Size is specified by two figures, swing and distance between centers. Together they specify the diameter and length of the largest cylinder that may be turned. Swing is twice the radial distance or *clearance* between the line connecting the center of the headstock and tailstock and the nearest point of the movable carriage.

There are many varieties of lathes and the names may describe a special operational capability or a degree of precision. The lathe illustrated in Fig. 2-1 is a 14-in swing by South Bend ®, one of several manufacturers and a specialist in the small- to medium-size range of 9 to 14 in (228 to 356 mm). Available for lathes of this size and type are many attachments that broaden the capability of the machine. Milling attachments, internal and external grinders, and others may be mounted on the compound and yield added versatility. These, plus an endless variety of chucks, centers, collets, and tools, allow a good machinist to make and repair many precise and complex machinery components. The lathe may be considered the most useful machine tool.

Proper cleaning and lubrication will preserve new *fits* for many years. Careful gear changing is a must since most of the speed and feed selection is achieved via spur gearing which does not lend itself to moving meshes. Safe operation always includes wearing safety glasses and avoiding loose clothing.

Figure 2-1 Typical maintenance lathe, 14-in swing South Bend®. *(South Bend Lathe, Inc., South Bend, Indiana).*

MILLING MACHINE

The milling machine matches the lathe in utility, and a medium-sized milling machine is a basic tool in all but the smallest shops. When parts cannot be procured in a timely manner, they must be made. When this happens, a milling machine is indispensable. Choose equipment with power capability of 1 to 4 hp.

The size of a mill is specified by the power available to the milling spindle, quill (or spindle) travel, collet capacity (diameter of the tool shank that can be held), table size, table travel in three dimensions (longitudinal or side-to-side, cross or front-to-back, vertical, or up-and-down, travel of the knee), spindle speeds available, and table and/or quill feeds available. Table travel under power feed may be different from travel under manual feed.

Figure 2-2 shows the Bridgeport ® Turret Miller, an optimum-size machine for the average facility. Many attachments have been developed for this machine by its maker and other tool companies since so great a number are in use by industry. Power table feed, optical or electronic dimension readout, slotting heads, and many other versatile attachments are available. While this particular machine is in wide use, there are several other good mills which are available for consideration.

A mill gives the capability of reproducing many broken or worn machinery parts. Except in the case of very accurate or special-purpose equipment, shop tools, if well cared for, will duplicate factory tolerances when skillfully operated within their capacity. The mills described offer diverse capability; vertical or horizontal milling plus angles between; boring, slotting, reaming, slitting, straddle milling, etc., via a wide range of tooling and attachments.

To use a milling machine properly requires a certain level of skill. The operator must first understand the use of basic cutting tools, i.e., end mills, shell mills, keyway cutters, side-cutting mills, etc. It is imperative that handbooks or the machine manual be con-

Figure 2-2 Typical maintenance milling machine, 2-hp Bridgeport® turret mill. *(Bridgeport Machines Division of Textron, Inc., Bridgeport, Connecticut).*

sulted to determine the feed, speed, and depth of cut that can be used within the power and rigidity capability of the milling machine. Bearing on this is the sharpness of the cutting tool and the geometry of the grind. An operator who is not thoroughly familiar with the operating controls of a particular machine should carefully study the manual or be given thorough training.

It is foolish to risk injury to either the person or the machine. Know exactly what movement will occur when a lever or handle is moved or engaged. A milled and defaced table is a symptom of lack of knowledge and attention. All tools are potentially dangerous; yet they are vital, valuable, and versatile. Keep your mill clean, free of chips, properly lubricated, and within its load capability and it will perform for a long time.

GAUGES AND OTHER MEASURING TOOLS

Most maintenance operations will depend on an appraisal of condition or degree of wear. This appraisal is best made by comparing *original* or *new* dimensions with existing or *worn* dimensions. The smallest shop requires precision scales, levels, vernier calipers, and micrometers. These will enable a wide variety of comparative checks on shaft diameters, thicknesses, lengths, and anything within the range of the instrument. Generally, we would be limited to a dimension approximating 24 to 36 in (0.61 to 0.91 m).

Measurement of the position or coordinates of machine components in different

planes or at distances beyond those above is extremely difficult, as is the similar operation of alignment verification, either during original setup or restorative operations.

Two excellent methods are available which do not depend on a *continuous* measurement device (one that extends from one surface to another). These are optical and laser
measuring and alignment techniques. Both depend on the intrinsic ability to establish
a reference or base line in space. This could be compared with placing widely separated
parts of a machine or assembly on a very large surface plate, except that the surface plate
reference can be defined in any plane. In effect, there is almost infinite flexibility in moving or placing this precise datum in space.

A combination example may illustrate the versatility of these techniques. Assume we
are to place 400 ft (122 m) of 12-in (304-mm) drainage pipe in a ditch with a continuous
slope of 1 in (25.4 mm) in 10 ft (3.04 m). (This is a minimum slope and hence, quite
critical.) A crude technique would be to place a 5-ft (1.52-m) section of concrete pipe,
establish an angle of slope by calculating the proportional rise to run, and then use a
carpenter's level or a line level to maintain this. More accurate would be the older optical
technique of the dumpy level or the more precise transit combined with the surveyor's
calibrated rod. After the rod is placed on each section, a *sighting* or *shooting* is taken, a
slope is calculated, and adjustments are made. This would establish a reasonably accurate slope compared to a perfectly level, horizontal plane. Alignment in a vertical plane
would have been more difficult. All the measurements might have been to the outside of
the pipe and subject to wall-thickness variation.

Imagine, instead, that you could slide each section of pipe over a perfectly straight
12-in (304-mm) diam rod, thereby realizing perfect alignment, and remove the rod when
alignment is complete. Lasers give this capability. The *gun* is centered via a *spider* (cen-
tering frame) in the pipe; the beam is properly inclined. As each section of pipe is laid,
a target spider is inserted in the end and the section is quickly adjusted till the beam
passes through the exact center of the target and therefore the pipe is perfectly aligned.
The technique is so simple as to be almost beyond belief! Each section is perfectly con-
centric to the desired axis, the rod, and the needlelike, nonscattered laser beam.

If your responsibility includes substantial erection and millwright work, investigate
both the older optical and the newer laser techniques.

No scale, level, plumb line, stretched wire, or the like can come close to the precision
and accuracy of either technique.

PORTABLE ELECTRIC INSTRUMENTS*

The variety of electric instruments grows daily. With the advent of solid-state circuitry,
sophisticated instruments once found only in a laboratory or test shop are now made in
hand-held varieties.

When selecting an electric instrument the most important features to look for are
perhaps reliability—will it take a beating and continue to operate; accuracy—is the read-
ing accurate within a reasonable deviation, say 2 percent; and functionality—does it do
the job for me?

There are so many new manufacturers of electric gear that it is difficult to sort out
the reputable brands. Be willing to consider some of the new equipment because it is
quite good. However, you could waste a lot of money on your own trials. Therefore, talk
to artisans, suppliers, and plant engineers.

Types of Portable Electric Instruments

The most common are discussed below.

*See also Chap. 3-7 of Rosaler and Rice: *Standard Handbook of Plant Engineering*, 1983.

Voltage Tester

A voltage tester is a handy pocket instrument with two probes. It is used to verify the presence of a voltage (a *live* circuit) and the approximate voltage level.

Multimeter

A multimeter is portable and battery operated; sometimes it is called a VOM, volt-ohm-milliammeter. It is used for more accurate measurements in diagnostic work, such as isolating high resistance, checking progressive voltage drop, etc., and is required in even the smallest shop.

Phase Tester

A phase tester is a portable unit which tells in which direction a three-phase motor will turn before it is connected to the line. It saves cut-and-try time and also identifies the phases for hookup or balancing purposes.

Clamp-On Ammeter

A clamp-on ammeter has loop jaws that separate to allow encircling of a current-carrying conductor without disconnecting it. The magnitude of the current is inductively measured and indicated on a meter. This tool is invaluable. It is now available in solid-state versions which also read voltage and resistance.

Megger Ohmmeter—Insulation Tester

This is another invaluable tool used to measure the breakdown resistance or resistance to ground of a current-carrying device. It is an excellent device for identifying grounded phase bars in a busway or for testing doubtful insulation on any electric device. It reads out in megohms. As a general rule the absolute minimum insulation resistance for voltages below 1000 is 1 MΩ or 1×10^6 Ω.

Others

There are numerous other instruments, but describing them all exceeds the scope of this book. Remember that by their nature these devices are somewhat fragile and should be treated with care. Meters of all types should be calibrated at least every 2 years. Some difficulties may be experienced with solid-state devices which do not have the inherent damping of the older moving-coil devices. They are often affected by spurious waveforms on a line and may yield erroneous readings. Users must be alert to this possibility in choosing an instrument.

BIBLIOGRAPHY

Scharff, Robert: *The Complete Book of Home Workshop Tools*, McGraw-Hill, New York, 1979.
Weygers, Alexander G.: *The Making of Tools*, Van Nostrand Reinhold, New York 1973.
Oberg, Erik and Franklin Jones: *Machinery's Handbook*, 18th ed. Industrial Press, New York 1969.
Carroll, Grady: *Industrial Instrument Servicing Handbook*, McGraw-Hill, New York, 1960.

chapter 1-3

Welding, Cutting, Brazing, and Soldering

by
Howard Cary
Vice President, Welding Systems
Hobart Brothers Co.
President, Hobart Brothers Technical Center
Troy, Ohio

GLOSSARY*

Arc blow   Magnetic disturbance of the arc which causes it to waver from its intended path.

Arc length   The distance from the end of the electrode to the point where the arc makes contact with the work surface.

Arc voltage   The voltage across the welding arc. It is measured with a voltmeter.

As-welded   The condition of the weld metal, welded joints, and weldments after welding and prior to any subsequent aging or thermal, mechanical, or chemical treatments.

Backhand welding   A welding technique in which the welding torch or gun is directed opposite to the progress of welding. It is sometimes referred to as the *pull gun technique* in gas metal arc welding and flux-cored arc welding. See travel angle, work angle, and drag angle.

Backing   Material (metal, asbestos, carbon, granulated flux, etc.) backing up the joint during welding.

Back-step welding   A welding technique wherein the increments of welding are deposited opposite the direction of progression.

Base metal   The metal to be welded, soldered, or cut.

Braze   A weld in which coalescence is produced by heating to a suitable temperature and by using a filler metal, having a liquidus above 800°F (427°C) and below the solidus of the base metals. The filler metal is distributed between the closely fitted surfaces of the joint by capillary attraction.

Butt weld   A weld made in the joint between two pieces of metal approximately in the same place. See Fig. 3-2.

Carbon steel   Carbon steel is a term applied to a broad range of material containing carbon, 1.7 percent max.; manganese, 1.65 percent max.; and silicon, 0.60 percent max. See the following table for classifications.

*Refer to Fig. 3-3 for a useful illustration of definitions of common welding nomenclature.

Type of Steel	Carbon Content, %
Low-carbon	0.15 max.
Mild-carbon	0.15–0.29
Medium-carbon	0.30–0.59
High-carbon	0.60–1.70

Cast iron A wide variety of iron-base materials—containing carbon, 1.7 to 4.5 percent; silicon, 0.5 to 3 percent; manganese, 0.2 to 1.3 percent; phosphorus, 0.8 percent max.; sulfur, 0.2 percent max.; molybdenum, nickel, chromium, and copper—can be added to produce alloyed cast irons.

Covered electrode A filler metal electrode, used in arc welding, consisting of a metal core wire with a relatively thick covering which provides protection for the molten metal from the atmosphere, improves the properties of the weld metal and stabilizes the arc.

Crater A depression at the termination of a weld bead or in the weld pool beneath the electrode.

Depth of fusion The depth of fusion of a groove weld is the distance from the surface of the base metal to that point within the joints at which fusion ceases. See Fig. 3-3.

Direct current Electric current which flows only in one direction. In welding it is an arc-welding process in which the power supply at the arc is direct current. It is measured by an ammeter.

Downhill welding A pipe-welding term indicating that the weld progresses from the top of the pipe to the bottom of the pipe. The pipe is not rotated.

Drag (thermal cutting) The offset distance between the actual and the theoretical exit points of the cutting oxygen stream measured on the exit surface of the material.

Elongation Extension produced between two gauge marks during a tensile test. It is expressed as a percentage of the original gauge length which should also be given.

Face of a weld The exposed surface of a weld, on the side from which welding was done. See Fig. 3-3.

Fillet weld A weld of approximately triangular cross section joining two surfaces approximately at right angles to each other in a lap joint, T joint, or corner joint. See Fig. 3-3.

Flat position The position in which welding is performed from the upper side of the joint and the face of the weld is approximately horizontal—sometimes called *downhand welding*. See Fig. 3-4.

Flux A fusible material used to dissolve and/or prevent the formation of oxides, nitrides, or other undesirable inclusions formed in welding which might contaminate the weld.

Flashback A recession of the flame into or back of the mixing chamber of the torch.

Flux-cored arc welding (FCAW) An arc-welding process in which coalescence is produced by heating with an arc between a continuous filler-metal (consumable) electrode and the work. Shielding is obtained from a flux contained within the electrode. Additional shielding may or may not be obtained from an externally supplied gas or gas mixture.

Gas metal arc welding (GMAW) (MIG) An arc welding process in which coalescence is produced by heating with an arc between a continuous filler-metal (consumable) electrode and the work. Shielding is obtained entirely from an externally supplied gas or gas mixture.

Gas-shielded arc welding See MIG and TIG welding.

Gas tungsten arc welding (GTAW) (TIG) An arc welding process wherein coalescence is produced by heating with an arc between a single tungsten (nonconsumable) electrode and the work. Shielding is obtained from a gas (argon or helium or a mixture of them). Pressure may or may not be used and filler metal may or may not be used.

Groove weld   A weld made in the groove between two members to be joined. See Fig. 3-3.

Heat-affected zone (HAZ)   That portion of the base metal which has not been melted, but the mechanical or microstructure properties of which have been altered by the heat of welding or cutting.

Horizontal position   See Fig. 3-4.

Impact resistance   Energy absorbed during breakage by impact of a specially prepared notched specimen, the result being commonly expressed in foot-pounds.

Kerf   The width of the cut produced during a cutting process.

Lap joint   A joint between two overlapping members.

Leg of a fillet weld   The distance from the root of the joint to the toe of the fillet weld. See Fig. 3-3.

Low-alloy steel   Low-alloy steels are those containing a low percentage of alloying elements.

Melting rate   The weight or length of electrode melted in a unit of time.

MIG welding   See gas metal arc welding.

Open-circuit voltage   The voltage between the output terminals of the welding machine when no current is flowing in the welding circuit. It is measured by a voltmeter.

Overhead position   The position of welding wherein welding is performed from the underside of the joint. See Fig. 3-4.

Overlap   Protrusion of weld metal beyond the bond at the toe of the weld.

Pass   A single longitudinal progression of a welding operation along a joint of weld deposit. The result of a pass is a weld bead.

Peening   Mechanical working of metal by means of impact blows with a hammer or power tool.

Penetration   The distance the fusion zone extends below the surface of the part(s) being welded.

Porosity   Gas pockets or voids in metal.

Post heating   Heat applied to the work after a welding, brazing, soldering, or cutting operation.

Preheating   The heat applied to the work prior to the welding, brazing, soldering, or cutting operation.

Pool   That portion of a weld that is molten at the place the heat is applied.

Radiography   The use of radiant energy in the form of x-rays or gamma rays for the nondestructive examination of metals.

Reduction of area   The difference between the original cross-sectional area and that of the smallest area at the point of rupture; usually stated as a percentage of the original area.

Reversed polarity   The arrangement of arc-welding leads in which the work is the negative pole and the electrode is the positive pole in the arc circuit. Abbreviated as DCEP.

Rheostat   A variable resistor which has one fixed terminal and a movable contact (often erroneously referred to as a "two-terminal potentiometer"). Potentiometers may be used as rheostats, but a rheostat cannot be used as a potentiometer because connections cannot be made to both ends of the resistance element.

Root of a weld The points, as shown in cross section, at which the bottom of the weld intersects the base metal surfaces. See Fig. 3-3.

Root opening The separation between the members to be joined at the root of the joint.

Shielded-metal arc welding An arc-welding process wherein coalescence is produced by heating with an electric arc between a covered metal electrode and the work. Shielding is obtained from decomposition of the electrode covering. Pressure is not used and filler metal is obtained from the electrode.

Silver solder A term erroneously used to denote silver-base brazing filler metal.

Size of a weld *Groove weld:* The joint penetration (depth of chamfering plus the root penetration when specified). *Fillet weld:* For equal fillet welds, the leg length of the largest isosceles right triangle which can be inscribed within the fillet-weld cross section. For unequal fillet welds, the leg lengths of the largest right triangle which can be inscribed within the fillet-weld cross section. See Fig. 3-3.

Slag inclusion Nonmetallic solid material entrapped in weld metal or between the weld metal and base metal.

Spatter In arc and gas welding, the metal particles expelled during welding and which do not form a part of the weld.

Stick welding See shielded-metal arc welding.

Straight polarity The arrangement of arc-welding leads in which the work is the positive pole and the electrode is the negative pole of the arc circuit. Abbreviated as DCEN.

Stress relief, heat treatment The uniform heating of structures to a sufficient temperature below the critical range to relieve the major portion of the residual stresses followed by uniform cooling.

Stringer bead A type of weld bead made without appreciable transverse oscillation.

Tack weld A weld (generally short) made to hold parts of a weldment in proper alignment until the final welds are made. Used for assembly purposes only.

Tensile strength The maximum load per unit of original cross-sectional area obtained before rupture of a tensile specimen. Measured in pounds per square inch.

Throat of a fillet weld Shortest distance from the root of a fillet weld to the face. See Fig. 3-3.

TIG welding See gas tungsten arc welding.

Toe of a weld The junction between the face of the weld and the base metal. See Fig. 3-3.

Tungsten electrode A nonfiller metal electrode, used in arc welding, consisting of a tungsten wire.

Ultimate tensile strength The maximum tensile stress which will cause a material to break. Usually expressed in pounds per square inch.

Underbead crack A crack in the heat-affected zone not extending to the surface of the base metal.

Undercut A groove melted into the base metal adjacent to the toe of the weld and left unfilled by weld metal.

Uphill welding A pipe-welding term indicating that the welds are made from the bottom of the pipe to the top of the pipe. The pipe is not rotated.

Vertical position The position of welding in which the axis of the weld is approximately vertical. See Fig. 3-4.

Weaving A technique of depositing weld metal in which the electrode is oscillated.

Weld A localized coalescence of metal in which coalescence is produced by heating to suitable temperatures, with or without the application of pressure and with or without the use of filler metal. The filler metal either has a melting point approximately the same as the base metal or has a melting point below that of the base metal but above 800°F (427°C).

Weld metal   That portion of a weld which has been melted during welding.

Welding procedure   The detailed methods and practices including joint welding procedures involved in the production of a weldment using a specific process.

Welding rod   Filler metal, in wire or rod form, used in gas welding and brazing processes, and those arc-welding processes in which the electrode does not furnish the filler metal.

Weldment   An assembly, the component parts of which are joined by welding.

Whipping   A term applied to an inward and upward movement of the electrode which is employed in vertical welding to avoid undercut.

BASIC PRINCIPLES

Welding is the most economical and efficient way to permanently join metal. It is the only way of joining two or more pieces of metal to make them act as a single piece. Welding is used to join almost all the commercial metals. However, some metals are more difficult to weld than others. There are more than 100 different welding processes and process variations which include brazing, soldering, and thermal cutting. These are broken into seven groups of welding processes and six groups of allied processes, as shown by the master chart of welding and allied processes, Fig. 3-1.

Many of the processes can be applied in different ways; i.e., they may be applied as manual, semiautomatic, machine, or fully automatic processes. Manual-process applications require a high degree of skill, while automatic welding requires a minimum amount of welding skill. This is important with respect to training and qualification of personnel.

There are certain fundamentals which must be grasped if one is to properly understand welding. Some of the most important are as follows:

The *weld joint* relates to the junction arrangement of the members to be joined. There are five basic types of joints, similar to those used by other crafts. These are: the butt joint, the corner joint, the edge joint, the lap joint, and the T joint. They are sometimes used in combination.

Another basic concept relates to the *type of weld*. See Fig. 3-2. The weld is the localized coalescence of metal at the specific junction of the parts. The many different types of welds are best described by the shape they show in cross section. Most popular are the *fillet* welds, followed by the *groove* weld. There are seven basic types of groove welds: square groove, bevel groove, V groove, J groove, U groove, flare V, and flare bevel. Many of these can be used in combination as double-groove welds. In order to completely describe a weld joint, the weld and the joint should both be defined (a "single V-welded butt joint," for example). See Fig. 3-3.

Some welds use *filler metals*. In the resistance-welding processes filler materials are normally not used. In the arc-welding processes filler metals are usually used. When filler metal is used, it must be properly specified in order to produce a weld joint of the specified strength.

The various *welding positions* (see Fig. 3-4) are particularly important when the welder's skill is involved. There are four basic positions: flat, horizontal, overhead, and vertical. These positions are rather obvious, but specific definitions are used. In identifying welding procedures, the position of the weld is highly important.

Another important factor is the *type of process:* electric or some other. The arc-welding processes are all electrically related, as are the resistance-welding processes. For gas welding, heat is obtained by chemical reactions of one type or another. In other cases, pressure is used, and it can be applied in various ways. In general, it is best to refer to the master chart of processes and then to the specific process that is to be employed.

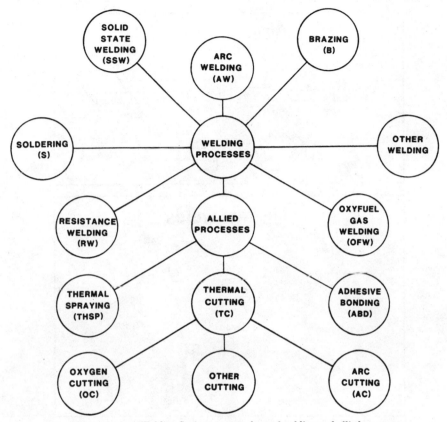

Figure 3-1   American Welding Society master chart of welding and allied processes.

Each process has its specific advantages and rationale for use. Many of them also have specific shortcomings and can not be used effectively in certain applications.

The safety and health of welders must always be considered. Welding is no more dangerous than other industrial occupations providing that the recommended safety precautions are followed. These precautions are given here and also appear on labels of filler metals and fluxes and on equipment for making welds.

This chapter describes the more popular welding processes in sufficient detail so that they can be properly understood and effectively applied. Included also are a glossary of the basic definitions of welding, a discussion of safety and health aspects, and guides to selecting filler material for welding various metals.

WELDING SAFETY AND HEALTH

These comments on safety are presented early in this chapter because it is important to understand the hazards involved *before* any welding operations are planned or begun.

Welding is no more hazardous than any other metalworking occupation providing that proper precautionary measures are followed. Hazards exist with any welding process. Welding is safe when safe practices are followed. The welder should follow safety precautions and supervisors must enforce safety regulations. The two most important

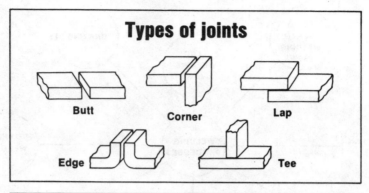

Types of joints

Butt   Corner   Lap

Edge   Tee

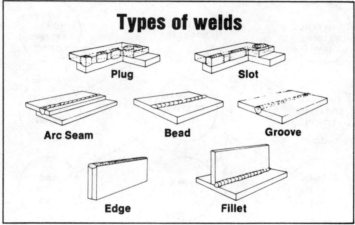

Types of welds

Plug   Slot

Arc Seam   Bead   Groove

Edge   Fillet

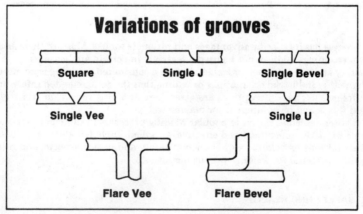

Variations of grooves

Square   Single J   Single Bevel

Single Vee   Single U

Flare Vee   Flare Bevel

Figure 3-2   Types of joints, welds, and grooves. *(American Welding Society.)*

GROOVE WELD

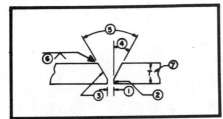

FILLET WELD

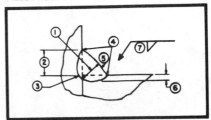

1. ROOT OPENING (RO): The separation between the members to be joined at the root of the joint.
2. ROOT FACE (RF): Groove face adjacent to the root of the joint.
3. GROOVE FACE: The surface of a member included in the groove.
4. BEVEL ANGLE (A): The angle formed between the prepared edge of a member and a plane perpendicular to the surface of the member.
5. GROOVE ANGLE (A): The total included angle of the groove between parts to be joined by a groove weld.
6. SIZE OF WELD(S): The joint penetration (depth of chamfering plus root penetration when specified).
7. PLATE THICKNESS (T): Thickness of plate welded.

1. THROAT OF A FILLET WELD: The shortest distance from the root of the fillet weld to its face.
2. LEG OF A FILLET WELD: The distance from the root of the joint to the toe of the fillet weld.
3. ROOT OF WELD: Deepest point of useful penetration in a fillet weld.
4. TOE OF A WELD: The junction between the face of a weld and the base metal.
5. FACE OF WELD: The exposed surface of a weld on the side from which the welding was done.
6. DEPTH OF FUSION: The distance that fusion extends into the base metal.
7. SIZE OF WELD(S): Leg length of the fillet.

Figure 3-3 Common terms applied to a weld. *(American Welding Society.)*

regulations concerning the subject are the American National Standard Z49.1, "Safety in Welding & Cutting," available from the American Welding Society and OSHA, "Safety & Health Standard 22CFR1910," available from the U.S. Department of Labor.

General Safety Rules

To protect yourself and others, read and understand these rules.

1. Electric shock
   a Electric shock can kill.
   b Do not touch live electric parts.
   c Make sure that the welding equipment is properly installed, the case is grounded, and the equipment is in good working condition.
   d Avoid welding in a wet or damp area. If this is unavoidable wear rubber boots and stand on a dry, insulated platform. Stay dry.
   e Always use insulated electrode holders. When not in use, hang the holder on brackets provided. Never place it under your arm.
   f Make sure that all electric connections are tight, clean, dry, and insulated.
   g Never attempt to repair electric equipment inside the welding machine or inside control panels, etc.
   h Make sure that power cables are insulated. Make sure that welding cables are insulated. Do not wrap cables around your body.
   i Do not use cables with frayed, cracked, or bare spots in the insulation. If there is a splice in the welding cable, make sure it is tight and insulated.
2. Arc radiation
   a Arc rays can injure eyes and burn skin.
   b Protect your eyes from the rays of the arc. Wear a head shield with the proper filter shade when welding or cutting. See a lens shade selector chart.

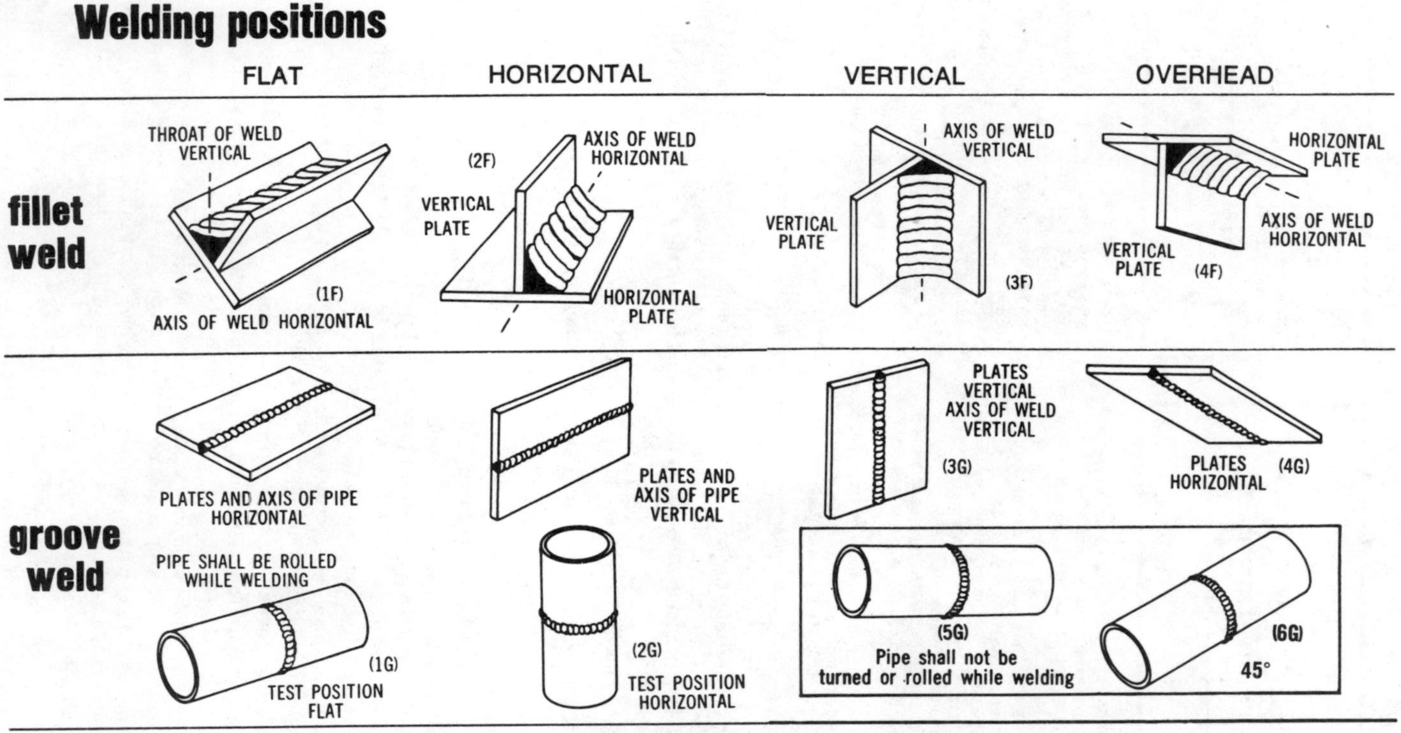

Welding positions

| | FLAT | HORIZONTAL | VERTICAL | OVERHEAD |

fillet weld

FLAT

THROAT OF WELD VERTICAL

(1F)

AXIS OF WELD HORIZONTAL

HORIZONTAL

AXIS OF WELD HORIZONTAL

(2F)

VERTICAL PLATE

HORIZONTAL PLATE

VERTICAL

AXIS OF WELD VERTICAL

(3F)

VERTICAL PLATE

OVERHEAD

HORIZONTAL PLATE

AXIS OF WELD HORIZONTAL

VERTICAL PLATE

(4F)

groove weld

FLAT

PLATES AND AXIS OF PIPE HORIZONTAL

PIPE SHALL BE ROLLED WHILE WELDING

(1G)

TEST POSITION FLAT

HORIZONTAL

PLATES AND AXIS OF PIPE VERTICAL

(2G)

TEST POSITION HORIZONTAL

VERTICAL

PLATES VERTICAL AXIS OF WELD VERTICAL

(3G)

(5G)

Pipe shall not be turned or rolled while welding

OVERHEAD

PLATES HORIZONTAL

(4G)

(6G)

45°

Figure 3-4   Welding positions. (*American Welding Society.*)

    c Be sure protective equipment is in good condition. Wear safety glasses in the work area at all times.

    d Wear protective clothing suitable for the welding work being done. Wear leather gloves and aprons with sleeves for heavy-duty welding. Protective clothing should shield the skin from arc rays.

    e Do not weld near degreasing operations. Arc rays may turn vapors into dangerous fumes.

    f Protect others from arc rays or flash with protective screens or barriers painted with nonreflecting paint.

3. Air contamination

    a Fumes and gases can be dangerous to your health.

    b Keep your head out of the fumes. Do not get too close to the arc.

    c Use enough ventilation and/or exhaust at the arc to keep fumes and gases from your breathing zone. Use natural drafts or fans to keep fumes away from your face.

    d Use mechanical exhaust when welding lead, cadmium, chromium, manganese, beryllium, bronze, zinc, or galvanized steel.

    e Do not weld in confined spaces without extra precautions.

    f Do not weld on plated materials or material covered with vinyl or heavy paint without mechanical exhaust. The coatings can release toxic fumes or gases.

    g Read and obey the warning label that appears on all containers of welding materials.

4. Fire and explosion

    a Arc welding and flame cutting involve high-temperature arcs and open flames which can create fires.

    b Keep your work area neat, clean, dry, and free of hazards.

    c Have fire-fighting equipment ready for immediate use and know how to use it.

    d Do not weld near flammable, volatile, or explosive liquids or gases. Remove all potential fire hazards from welding area.

    e Do not weld on or near fuel tanks of engine-driven equipment.

    f Do not weld on containers such as drums, barrels, or tanks that may have held combustibles or hazardous materials without taking extra special precautions. See the AWS Bulletin, "Safe Practices for Welding and Cutting Containers That Have Held Combustibles."

    g Do not weld on sealed containers or compartments without providing vents and taking extra precautions.

5. Compressed gases

    a Handle all compressed-gas cylinders with extreme care. Keep cylinder caps on when cylinder is not in use.

    b Make sure that all gas cylinders are secured to the wall or other structural support. Protect them from mechanical shocks.

    c *Never* strike an arc on a compressed-gas cylinder. A gas cylinder should *not* be a part of an electric circuit.

    d When compressed-gas cylinders are empty, close the valve and mark the cylinder "EMPTY."

    e Store compressed-gas cylinders in a safe place with good ventilation. Acetylene cylinders and other fuel-gas cylinders should be stored separately from oxygen cylinders. Avoid excessive heat.

    f Acetylene cylinders should be stored and used in the vertical position.

6. Cleaning and chipping welds and other hazards

    a Wear protective chipping goggles when chipping weld slag. Chip away from your face.

b When you are grinding or using power tools, you should wear safety glasses with side shields under the welding helmet.

c Dispose of electrode stubs in containers; stubs on the floor are a safety hazard.

d When working above ground make sure that scaffolds, ladders, or work surfaces are substantial and solid.

e When welding in high places without railings, use a safety belt or lifeline.

f When working in noisy areas or using noisy processes, wear ear protection.

g When working in confined areas take special precautions because of fire and explosion problems with fuels. Guard against inert gas or fume buildup from welding. Provide lookouts and special ventilation.

ARC WELDING AND CUTTING PROCESSES

The Shielded-Metal–Arc-Welding Process

Process

Shielded-metal arc welding (SMAW), sometimes called stick welding or manual metal arc welding, is the most popular welding process in use today (Fig. 3-5). It is an electrical

Figure 3-5 Application of shielded-metal arc welding. *(Hobart Brothers Company.)*

arc-welding process which fuses the parts to be welded by heating them with an arc between a covered consumable metal electrode and the work. Shielding is obtained from the decomposition of the electrode covering. This process became very popular in the early 1930s when the different kinds of coatings were developed. SMAW is normally manually applied and can be used for welding thin and thick steels and some nonferrous metals in all positions. The process requires a relatively high degree of welder skill. The diagram of the SMAW process in Fig. 3-6, shows the covered electrode, the core wire, the arc, the shielding atmosphere, the weld, and the solidified slag. Deposited metal is obtained from the end of the electrode, which melts and crosses the arc.

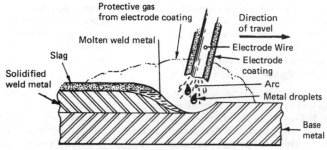

Figure 3-6   Process diagram for shielded-metal arc welding.

Application

This manually controlled process welds all nonferrous metals ranging in thickness from 18 gauge (0.048 in) to the maximum encountered. When material thicknesses are over ¼ in, a bevel edge preparation is used and a multipass welding technique is employed. The process allows for all-position welding. The arc is under the control of, and is visible to, the welder. Slag removal is required.

Equipment

The major parts needed for the SMAW process are: (1) the welding machine (power source), (2) the covered electrode, (3) electrode holder, and (4) welding leads or cables to complete the welding circuit. These are shown in Fig. 3-7.

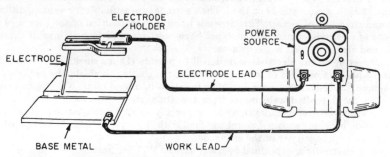

Figure 3-7   Equipment for shielded-metal arc welding.

Welding Machine.   The welding machine (power source) is the most important item of welding equipment required. It must provide electric power of the proper current and voltage sufficient to maintain a stable welding arc. The SMAW process can be used with either alternating or direct current, the direct current being of either polarity. Straight polarity is with the electrode negative; reverse polarity is with the electrode positive.

There are many different types of welding machines. The least expensive, lightest weight, and smallest welding machine is the alternating-current (ac) transformer type. It provides alternating welding current at the arc. It is usually a single-control type of machine having one knob which is used to vary the current output. Other types have plug-in connectors or tap switches for this purpose. Transformer machines range from the smallest hobby type up to heavy industrial machines for automatic welding.

The rectifier-type welding machine converts ac power to dc power and provides direct current at the arc. This type of machine usually has a single control knob; however, range switches are sometimes used. These machines come in various sizes.

Another power source is the ac-dc transformer-rectifier type that is especially designed to allow either ac or dc welding. A selector switch allows either alternating current or direct current of either polarity.

Another type of power source for welding is the dual-control dc generator. This type of machine allows for adjustment of the open-circuit voltage and output slope as well as the welding current. Where electric power is available, the generator is driven by an electric motor. Away from the power line, the generator will be driven by an internal-combustion engine fueled by either gasoline or diesel oil. Belt-driven generators with a power takeoff are also available.

Electrode Holder. The electrode holder, which is held by the welder, is used to grip the electrode and carry the welding current to it. Only electrically insulated holders should be used. They come in different types, primarily the pincher type and the twist-collet type. They also come in different sizes rated according to the maximum current that can be used. Holders having larger current ratings are heavier and will also accommodate larger cables. Personal preference of the welder has much to do with the selection of electrode holders.

Welding Leads. The welding cables and connectors provide the electric circuit necessary to transmit power from the welding machines to the arc. The *electrode lead* forms one side of the circuit and runs from the electrode holder to the electrode terminal of the power source. The *work lead* (erroneously called ground lead) is the other side of the circuit and runs from the work clamp to the work terminal of the welding machine. Welding cables are made of many strands of copper wire; aluminum, however, is sometimes used. The cable is covered by a sheath of tough insulating material to protect it and avoid short circuits. The cable size is based on the welding current to be used. Cable sizes range from AWG No. 6 to AWG No. 4/10, which is the largest and is used for heavy-duty applications. The leads should be no longer than is required for the work to be done.

Covered Electrodes. Covered electrodes come in various diameters from $\frac{1}{16}$ to $\frac{1}{4}$ in, and their length is normally 14 or 18 in. The electrodes are available for welding different types and strengths of metals. This depends upon the composition of the core wire and the type of electrode coating. Other types have been designed to match most common metals and also to deposit hard surfaces.

The covering on the electrode is designed to provide (1) gas shielding—obtained by the decomposition of some of the ingredients in the coating to shield the arc from the atmosphere, (2) deoxidizers—for purifying the deposited weld metal, (3) slag formers—to protect the deposited weld metal from the atmosphere, (4) ionizing elements—to make the electrode operate more smoothly, especially on alternating current, (5) alloying elements—to provide deposited metal matched to the base metal, and (6) iron powder—to improve the productivity of the electrode.

Covered electrodes are specified by the American Welding Society Specification A5.1, "Carbon Steel Covered Arc Welding Electrodes," and A5.5, "Low Alloy Steel Covered Electrodes," as well as others for the different electrodes available.

Gas Tungsten Arc-Welding Process

Process

Gas tungsten arc welding (GTAW), also known as TIG welding, heliarc welding, heli welding and argon arc welding is one of the newer welding processes. It is illustrated in Fig. 3-8. It is an electric arc welding process which fuses the parts to be welded by heating them with an arc between a nonconsumable tungsten electrode and the work. Filler metal may or may not be used. Shielding is obtained from an inert gas or an inert gas mixture. The process is normally applied manually and is capable of welding steels and nonferrous metals in all positions. The process is commonly used to weld thin metals and for the root-pass welding on tubing and pipe. It requires a relatively high degree of welder skill and produces excellent quality welds.

GTAW was developed by the aircraft industry in the early 1940s to join hard-to-weld metals, particularly magnesium, aluminum, and stainless steels. Figure 3-9 is a diagram of the GTAW process. The tungsten electrode is fastened in a torch which also has a

Figure 3-8   Application of gas tungsten arc welding. *(Hobart Brothers Company.)*

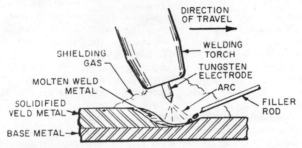

Figure 3-9   Process diagram for gas tungsten arc welding.

nozzle for directing the shielding gas around the arc area. The arc is between the tungsten electrode and the work. Filler metal in the form of a rod or wire is usually fed manually although it can be fed automatically.

Application

The outstanding features of the GTAW process are (1) the ability to produce high-quality welds on almost all metals and alloys; (2) little or no postweld cleaning is required; (3) the arc and weld pool are clearly visible to the welder; (4) there is no filler metal crossing the arc, hence little or no weld spatter; (5) welding is possible in all positions; (6) there is no production of slag which might be trapped in the weld. GTAW can be used for welding aluminum, magnesium, stainless steel, bronze, silver, copper and copper alloys, nickel and nickel alloys, cast iron, and steel. It will weld a wide range of metal thicknesses but is most popular on thinner gauges. Argon is usually used as the shielding gas, although helium or argon-helium mixtures are sometimes used.

Equipment

The major components required for GTAW are shown in Fig. 3-10. These items are (1) the welding machine or power source, (2) the GTAW torch, including the tungsten electrode, (3) the shielding gas and controls, and (4) the filler rod when required. There are several optional accessories available including a remote-controlled foot rheostat which permits the welder to control current while welding; others are arc timers and controllers, high-frequency units, water circulating systems, and specialized devices.

Welding Machine.   A specially designed welding machine or power source is used for GTAW. In general, power sources for GTAW have drooping characteristics. Both alter-

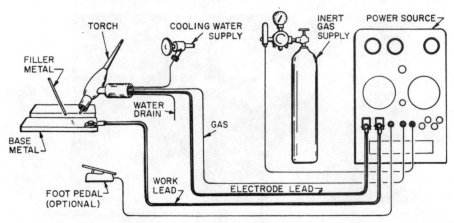

Figure 3-10   Equipment for gas tungsten arc welding.

nating and direct current are used. A transformer-type, transformer-rectifier-type, or generator-type power source can be employed. The power source usually contains a high-frequency generator which is used to aid arc starting when welding with direct current and it used continuously when welding with alternating current. The selection of alternating or direct current depends on the material being welded. Alternating current is recommended for welding aluminum and magnesium. Direct current is recommended for welding stainless steels, carbon steels, copper and its alloys, nickel and its alloys, and precious metals.

Most machines designed for GTAW include solenoid valves for controlling the shielding gas and cooling water, when used. The high-frequency spark gap oscillator is also included in the welding machine, as well as the special connectors for attaching the welding torch and cable assembly to the machine. The welding machine may also include meters and programmers. Some machines provide pulsed current capability.

Welding machines designed for GTAW can also be used for SMAW and several other processes.

It is possible to use conventional ac or dc power sources designed for SMAW. However, a high-frequency attachment is usually required, and the machine must be downrated when welding with alternating current. Best results are obtained with a welding machine specifically designed for GTAW.

Welding Torch.   The GTAW torch holds the tungsten electrode and directs the shielding gas and welding power to the arc. Torches come in different sizes, and the larger sizes are usually water-cooled. The torches normally come equipped with a cable assembly which directs the gases, welding-power current, and cooling water (when used) from the machine to the torch.

Tungsten Electrodes.   The electrodes used with the GTAW process are made of tungsten or tungsten alloys. Tungsten has the highest melting point of any metal (6170°F or 3405°C) and is considered nonconsumable. When properly used, the electrode does not touch the molten weld puddle. If the tungsten electrode accidentally touches the weld puddle, it becomes contaminated and must be cleaned immediately. If it is not cleaned, an erratic arc will result. Electrodes are available in three alloys as well as pure tungsten. Pure tungsten is the least expensive; however, alloys containing 1 or 2 percent thoriated tungsten are quite popular. This type of electrode is somewhat more expensive but is recommended for welding specific metals. Another alloy is the zirconiated tungsten which is often used for x-ray quality work. The different tungstens are identified by AWS Specification A5.12, "Specifications for Tungsten Arc Welding Electrodes," and are color-coded for ease of recognition. Tungsten electrodes come in diameters ranging from

0.020 in (.5 mm) up through ¼ in (6 mm). The electrode surfaces come in either a ground finish or a cleaned finish. The lengths of tungsten rods are normally 3 to 6 in (7.5 to 15 cm).

Shielding Gas.   An inert shielding gas must be used. Argon is the most popular; however, helium is used in certain applications, and in some cases a mixture of argon and helium is used. Argon is more easily obtainable and is less expensive than helium. Also it is heavier than helium and provides better shielding at lower flow rates. The arc produced with helium for shielding is considered hotter and is used for obtaining deeper penetration. Helium is also used for welding in the overhead position. Helium is usually used at a higher flow rate than argon.

Filler Metal.   Though filler metal may or may not be used, it is normally used except when one is welding very thin material. The composition of the filler metal should match that of the base metal. Filler metals may not be available in every alloy; therefore, filler-metal charts show the recommended type for use. The size of the filler-metal rod depends on the thickness of the base metal and the welding current. Filler metal is usually added manually to the weld puddle, but automatic feed is sometimes used. AWS specifications provide information about filler wires available.

Welding Safety

Welding safety for GTAW is essentially the same as for the other arc-welding processes; however, the process should not be used near chemical-cleaning tanks since the arc rays may change the gas vapors to poisonous vapors. Also, the filter glass shade in the helmet must be related to the proper welding current. Ventilation should be provided when one is welding in confined areas since argon, being relatively heavy, will tend to stay in the area and gradually displace the breathing air of the welder.

Gas Metal Arc-Welding Process

Process

The gas metal arc-welding process (GMAW) is an arc-welding process which fuses the parts to be welded by heating them with an arc between a continuous, consumable solid wire electrode and the work. Shielding is obtained from an externally supplied gas or gas mixture. The process is normally applied semiautomatically; however, it can be applied by machine or by automatic equipment. The process can be used to weld thin and fairly thick metals, both steel and nonferrous. The arc is visible to the welder, and it can be used in all positions. A lesser degree of welding skill is required; however, the equipment is more complex than that used for SMAW.

This process, shown in Fig. 3-11, is one of the newer arc welding processes. It was developed in the early 1950s and became extremely popular in the 1960s.

This process is sometimes called MIG welding (standing for metal–inert gas welding), or microwire welding, short-arc welding, dip transfer welding, CO_2 welding, etc. The electrode is melted in the heat of the arc, and the metal is transferred across the arc to become the deposited weld metal.

The GMAW process is also shown in Fig. 3-12. The illustration shows the electrode wire, the nozzle of the welding gun or torch, the shielding-gas envelope, and the arc between the end of the electrode and the base metal.

There are a number of variations of the GMAW process. These depend on the type of shielding gas which relates to the type of metal transfer across the arc as follows:

- MIG welding using inert-gas shielding on nonferrous metals.
- Short circuiting transfer (microwire) normally using CO_2 gas or CO_2 gas mixtures and small-diameter electrode wire allows welding in all positions and on thin metals.
- CO_2 welding using CO_2 shielding gas and larger electrode wires and is restricted to steels.

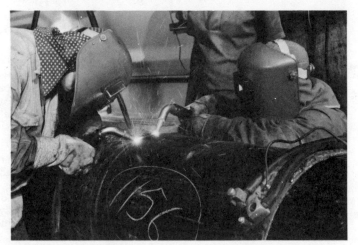

Figure 3-11 Application of gas metal arc welding. *(Hobart Brothers Company.)*

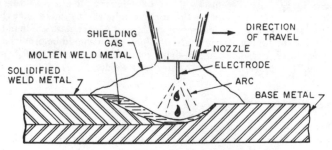

Figure 3-12 Process diagram for gas metal arc welding.

- Spray arc welding which uses the argon-oxygen shielding gas normally restricted to steels.
- Pulsed arc welding which provides pulsed metal transfer and uses a special power source.

Application

The outstanding features of GMAW are: (1) high-quality welds on most metals; (2) minimum postweld cleaning is required; (3) the arc and weld pool are visible to the welder; (4) welding is possible in all positions depending on electrode size; (5) relatively high-speed welding; (6) there is little or no slag produced; (7) it is considered a low-hydrogen-type welding process.

Variations of the process offer special advantages. The short-circuiting arc (micro-wire) will weld most steels in the thinner gauges. CO_2 welding allows for high-speed travel on steel. The spray arc variation produces high-speed welds with minimum spatter and cleanup, and the MIG process welds the nonferrous metals at a higher rate of speed than the GTAW process.

Equipment

Major components required for GMAW are shown in Fig. 3-13. These are (1) the welding machine or power source, (2) the electrode wire-feed system and control, (3) the welding gun and cable assembly (for semiautomatic welding or welding torch for automatic welding), (4) the shielding gas supply and controls, and (5) the consumable electrode.

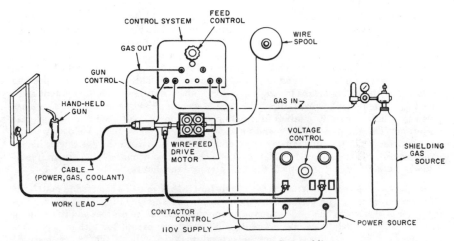

Figure 3-13   Equipment for gas metal arc welding.

Welding Machine.   The power source for GMAW is normally a constant-voltage (CV) or constant-potential (CP) type. Its characteristic output volt-ampere curve is essentially flat with a small droop. Thus the output voltage is approximately the same even though the welding current changes. The output voltage is adjusted at the power source which can be a transformer-rectifier, a motor-generator, or an engine-driven generator. A CV or CP power source does not have a welding-current control and is not used for the shielded-metal arc process. The welding-current output is determined by the electric load on the machine which depends on the electrode wire-feed speed. A dc electrode positive (DCEP) arrangement is normally used. Machines for this process are available from 150 A up to as high as 1000 A and should be rated at 80 to 100 percent duty cycle. They should include a contactor and meters and should provide 115-V ac power for the electrode wire feeder.

Wire Feeder.   The wire-feed system must be matched to the power supply. The CV system of welding relies on the relationship between the electrode wire burn-off rate and the welding current. This relationship is fairly constant for a given electrode wire size, composition, and shielding atmosphere. At a given wire-feed speed rate, the welding machine will supply the proper amount of current to maintain a steady arc. Thus the wire-feed speed control adjusts the welding current. The CV welding system is a self-regulating system and is recommended when using small-diameter electrode wires. A miniaturized wire feeder built into the welding gun is popular for welding with small-diameter aluminum wire. The wire-feed system and controls are essentially the same for semiautomatic, machine, or automatic welding.

Welding Gun.   The welding gun and cable assembly are used to carry the electrode wire, the welding current, and the shielding gas to the welding arc. For higher-current applications, water-cooled guns are used and the water is also carried through the cable assembly. There are two general types of welding guns, the pistol-grip and curved-head (gooseneck). The gooseneck type is more popular for small-diameter-electrode wire. The pistol-grip type is usually used for welding with larger electrode wires and for welding with nonferrous electrodes. Guns used for heavy-duty work at high currents and guns using inert gas for shielding at medium to high currents are water-cooled. For machine or automatic welding, a welding torch is used. The automatic torches are either air- or water-cooled depending on the welding application, as mentioned above. For CO_2 welding a side-delivery gas nozzle is often used with automatic torches. The wire guides in all guns and torches must match the size of the electrode wire being used.

Shielding Gas.   The shielding gas displaces the air around the arc to prevent contamination by the oxygen and nitrogen in the atmosphere. The gaseous shielding envelope must efficiently shield the arc area in order to obtain high-quality weld metal. Various shielding gases can be used depending on the process variation and the base metal being welded. Carbon dioxide is the least expensive and is very popular. Mixtures of CO_2 and argon and mixtures of argon and oxygen are also used. Shielding gas must be specified "welding grade." This means the gas has a high purity and low moisture content indicated by its dew-point temperature. The type of gas for shielding and the flow rate are given by welding procedure tables for welding various metals with the different process variations. The gas-flow rates depend on the type of gas, metal being welded, welding position, etc. When welding outside or when air currents disturb the gas shield, higher gas-flow rates are necessary. For high flow of CO_2 gas, two or more cylinders are manifolded together to avoid freezing of the CO_2 pressure regulators.

Electrode.   The composition of the electrode for GMAW must be selected to match the metal being welded, the variation of the process, and the shielding atmosphere. The diameter or size of the electrode depends on the variation of the process and the welding position. All electrode wires are normally solid and bare, except for a thin, protective coating on carbon steel wires. The welding procedure tables indicate the proper electrode wire, type, and size for welding different metals. Electrode wires are available in a wide variety of diameters, spools, coils, and reels and are specified by AWS specifications.

Flux-Cored Arc-Welding Process

Process

The flux-cored arc-welding process (FCAW), also known as FabCO®*, Dualshield, Fabshield®*, self-shield, Innershield, etc., is an arc-welding process which fuses the parts to be welded by heating them with an arc between a continuous flux-filled electrode wire and the work. Shielding is obtained from gas generated by the decomposition of the flux within the tubular wire; however, additional shielding may be obtained from an externally supplied gas or gas mixture. The process, usually applied semiautomatically, is shown in Fig. 3-14. It also can be applied by machine or automatic equipment. It is normally used for welding medium-thick steels and stainless steel and for surfacing. It is not normally used for welding nonferrous metals. Small-diameter electrodes enable the use of all positions of welding. With larger-size electrodes, the welder is restricted to the flat and horizontal positions. The arc is visible to the welder and the skill level required is similar to that for GMAW.

The process diagram shown in Fig. 3-15 shows the two variations, with the optional items for the externally gas-shielded variation indicated by the dotted lines. The flux-cored electrode wire and the arc between it and the base metal are shown. The process normally produces a slag covering which must be removed after welding. The externally gas-shielded variation was the original process and employed CO_2 for external shielding. The other, or self-shielding variation, generates sufficient shielding gas from the decomposition of the ingredients in the core of the electrode wire. In either case, the gas shield prevents the atmospheric oxygen and nitrogen from reaching the arc area. This process was developed in the mid-1950s and became popular in the 1960s.

Application

The two variations of the process provide slightly different welding features. With external shielding gas, the features of the process are (1) extremely smooth, sound, high-quality welds, (2) deep penetration, and (3) good properties for x-ray–quality welds.

The gasless or self-shielding variation offers the following features: (1) elimination of gas supply, controls, and gas nozzle, (2) moderate penetration, and (3) ability to weld in drafts or breezes.

*® Hobart Brothers Co.

Figure 3-14   Application of flux-cored arc welding. (*Hobart Brothers Company.*)

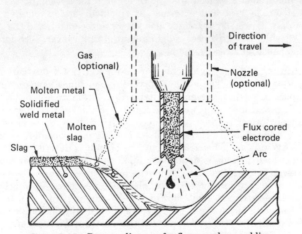

Figure 3-15   Process diagram for flux-cored arc welding.

Both variations have the following features: (1) high deposition rates, (2) visibility of the arc to the welder, (3) all position welding based on the size of the electrode, and (4) similarity of the weld-joint design to those used for the other arc-welding processes.

Both variations are normally restricted to the welding of carbon and stainless steels and for overlaying. The external gas-shielded version can be used for welding many low-alloy steels.

Equipment

The major components required for the FCAW process are shown in Fig. 3-16. Equipment is generally similar to that used for GMAW and is common for both variations

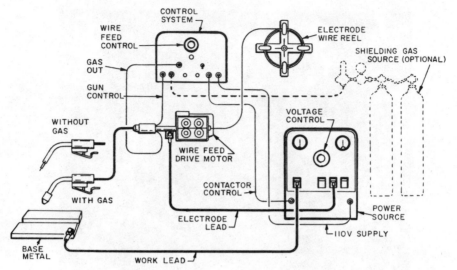

Figure 3-16   Equipment for flux-cored arc welding.

except for the gas shielding system. The items involved are (1) the welding machine or power source, (2) the wire-feed drive system and control, (3) the welding gun and cable assembly for semiautomatic welding or a welding torch for automatic welding, and (4) the flux-cored electrode wire. The external gas-shielded version requires the external shielding gas supply, flowmeter-regulator, gas valves and control, and the gas nozzle on the gun. The self-shielding type uses a lightweight gun; however, such guns often will include smoke-exhaust nozzles.

Welding Machine.   The welding machine or power source for flux-cored arc welding is normally a CV or CP type. These types of welding machines have an output characteristic volt-ampere curve that is essentially flat with a minimum droop. The output voltage for the welding machine is adjusted by a control on the welding power source which can be either a transformer-rectifier or a generator driven by a motor or an engine. CV-type power sources do not have a current control and, therefore, cannot be used for welding with the SMAW process. The welding-current output is determined by the electric load on the power source. This is dependent upon the electrode wire-feed speed rate. Direct-current electrode positive (DCEP) is the arrangement normally used; however, some electrodes use direct current with the electrode negative. Alternating current is used rarely. Power sources are available for FCAW ranging from 150 to 1000 A and should be rated at 80 to 100 percent duty cycle. They should include a contactor and meters, and should provide 115-V-ac power for the electrode wire feeder.

Wire Feeder.   The wire-feeding mechanism feeds the flux-cored electrode wire automatically from a coil or spool to the cable assembly and welding gun into the arc. The wire-feed system must match the type of power supply used. The CV-type power supply is normally used; therefore, a constant-speed wire-feed system with adjustable speed is used. The wire-feed speed rate controls the welding current. The CV welding system is a self-regulating system. Voltage-sensing wire-feed systems can be used when matched to a drooping-characteristic-type power source, but they are not too popular for FCAW. Basically the same type of wire feeder that is used for GMAW can be used for FCAW.

Welding Gun.   The welding gun is used to deliver the electrode wire, the current, and the shielding gas (when used) to the arc area. Guns with shielding gas nozzles are water-cooled for high current, 500 A or more, heavy-duty cycle welding. Water cooling is not used for the gasless variation welding gun. Both pistol-grip and gooseneck guns are available. Sometimes with the gasless variation, a special insulated extension which adds to the electrical stickout is added to the gun to provide higher deposition rates.

Shielding Gas (External Gas-Shielded Variation).   The shielding gas displaces the air around the arc area, preventing contamination by oxygen and nitrogen of the atmosphere. CO_2 is normally used as the shielding gas for steel; however, for stainless steel and certain alloy steels, a gas mixture is used. The type of shielding gas must be related to the electrode wire and base metal. Gas-flow rates depend on the type of gas being used, the metal being welded, welding position, welding current, etc. Procedure tables provide this information.

Electrode Wire.   The electrode wire employed must be selected to match the composition and mechanical properties of the base metal. The selection must also be based on whether it is to be used with external shielding gas or not. Procedure tables usually indicate the type of electrode wire to be used. Various diameters are available for different applications. Electrode wires are packaged on spools, coils, and in payoff-type packs. The American Welding Society classifies flux-cored electrodes according to the strength level, properties, and deposited weld metal composition.

Submerged Arc-Welding Process

Process
Submerged arc welding (SAW), also known as welding under powder, hidden arc welding and union melt welding, is an arc-welding process which fuses the parts to be welded by heating them with an arc or arcs between a bare electrode or electrodes and the work. The arc is shielded by a blanket of granular flux on the work. The process is normally applied by machine or automatically but is used on a limited basis semiautomatically. It is used to weld medium to thick steels in a flat or horizontal position. Manual welding skill is not required; however, a technical understanding of the equipment and the welding procedure is necessary. SAW, shown in Fig. 3-17, was developed in 1930 by the

Figure 3-17   Application of submerged arc welding. (*Hobart Brothers Company.*)

National Tube Company to make longitudinal welds in pipe. It has become extremely popular for heavy plate welding because it produces high-quality weld metal at a minimum cost. Figure 3-18 shows the base metal, the consumable electrode wire, the granular

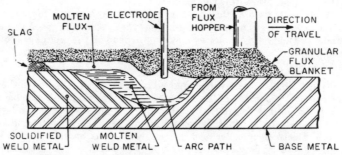

Figure 3-18 Process diagram for submerged arc welding.

flux covering, the slag cover, the arc area, and the molten metal. SAW is normally used for welding steels and is not used for welding nonferrous metals.

Application

The outstanding features of the SAW process are (1) high metal deposition rates, (2) high welding travel speed, (3) deep penetration, (4) good x-ray quality, (5) smooth weld appearance, (6) easily removed slag covering, and (7) a wide range of weldable metal thicknesses. The arc is not visible to the welder. The automatic or machine methods of application are most commonly used. The semiautomatic application method is less popular. SAW is used to weld low- and medium-carbon steels, low-alloy high-strength steels, quenched and tempered steels, and many stainless steels. It is also used for hardsurfacing, hardfacing, and buildup work. Metal thicknesses ranging from 16 gauge to ½ in are welded with no edge preparation. With edge preparation and multiple-pass welding, the maximum thickness welded is practically unlimited. SAW is restricted to the flat and horizontal positions.

Equipment

The major equipment components required for SAW are shown in Fig. 3-19. These are (1) the welding machine (power source), (2) the wire-feeding mechanism and control, (3) the welding torch for automatic welding or the welding gun and cable assembly for semiautomatic welding, (4) the flux hopper and flux feeding mechanism, and (5) the travel mechanism for automatic welding. A flux recovery system is usually included in an automatic installation.

Welding Machine. The welding machine or power source for SAW can be either an ac or dc power source. It must be rated at a 100 percent duty cycle since welding operations are continuous and the length of time in operation will normally exceed the 10-min base period used for rating duty cycle. For dc SAW, the CV-type or CC-type power source can be used. The CV type is more common for small-diameter electrode wires, usually ⅛ in and smaller in diameter. The CC type is more commonly used for larger-diameter electrode wires, usually ⁵⁄₃₂ in and larger. The wire feeder must be matched to the type of power source used. When alternating current is employed, the machine is a CC type. Welding machines for SAW range in size from 200 to 1000 A. In some cases two or more electrode wires are employed in the same puddle, and in other cases one electrode may be on direct current and the other on alternating current.

Wire Feeder. The wire-feeding mechanism and its associated control feed the electrode wire into the welding arc. When a CC or drooping-type power source is employed,

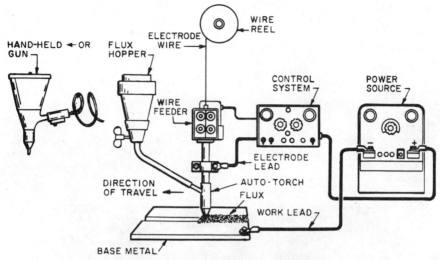

Figure 3-19   Equipment for submerged arc welding.

a voltage-sensing wire-feeder system must be used. This type of wire feeder maintains a specific arc voltage and feeds the electrode wire at the proper rate to maintain this value. If a CV or flat-characteristic power source is used, the constant-speed wire feeder and control should be employed. In this case, the wire feeder feeds the electrode wire at a constant but adjustable rate in order to draw the prescribed welding current from the power source. The arc voltage is adjusted by changing the output voltage of the power source. The control system initiates the arc, provides the proper electrode wire-feed speed and, in automatic operation, performs other necessary functions such as start and stop of fixture travel.

Welding Torch or Gun.   For automatic welding the torch directs the electrode wire into the arc and transfers the welding current to the wire as it leaves the torch. For automatic welding, the torch is usually attached to the electrode wire-feeder and travel mechanism. A flux hopper is usually attached to or is adjacent to the torch. For semiautomatic operations a welding gun and cable assembly are used to transmit the electrode wire and the welding current to the arc and to provide the flux at the welding zone. A small flux hopper may be attached to the gun, and it dispenses flux over the weld area in accordance with the manipulation of the gun. In another system, the flux is fed through a conduit to the gun from a hopper and is dispensed at the welding zone. Semiautomatic guns usually have a trigger switch for initiating the arc.

Welding Flux.   The SAW flux is a granular, fusible material which is poured over the arc area. This flux performs the same functions as the covering on a coated electrode. It protects the arc and molten metal from atmospheric contamination, acting as a scavenger to clean and purify the weld metal. Additionally, it may be used to add alloy elements to the deposited weld metal. A portion of the flux is melted by the heat of the welding arc. The molten flux then cools and solidifies, forming a slag on the surface of the weld. The portion of the flux which is not melted can be recovered and reused. There are different grades and types of submerged arc flux and it is important to select the proper flux-wire combination to match the chemistry and properties of the metal being welded. AWS Specification 5.17 provides the information necessary to match the properties of the metal being welded.

Electrode.   The electrode wires used for SAW are usually solid and bare except for a thin, protective coating on the surface, usually copper. The electrode contains deoxidizers which help clean and scavenge the weld metal to produce a quality weld. Alloying elements may also be included in the composition of the electrode. The electrode composition and the type of flux must be matched to the requirements of the base metal in order to provide a quality weld. This is covered by AWS Specification 5.17. Electrode wires are available in diameter sizes of ¹⁄₁₆, ⁵⁄₆₄, ³⁄₃₂, ⅛, ⁵⁄₁₆, ⁷⁄₃₂, and ¼ in. Wire is usually available in coils ranging from 50 to 1000 lb.

The Electroslag Welding Process Consumable Guide System

Process

Electroslag welding (ESW), also known as Porta-Slag* or slag welding, is a welding process that fuses the parts to be welded with molten slag which melts the filler metal and the surfaces of the work to be welded. The molten weld pool is shielded by a slag covering which moves along the joint as welding progresses. The process, shown Fig. 3-20, is not

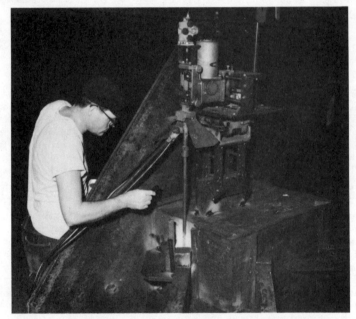

Figure 3-20   Application of electroslag welding.(*Hobart Brothers Company.*)

an arc-welding process, except that an arc is used to start the process. After stabilization the molten slag provides the necessary heat for welding. The process is applied automatically. It is a limited-application process used for making vertical welds on medium to heavy thicknesses of steel. Manual welding skill is not required, but a technical knowledge of the process is necessary to operate the equipment.

A diagram of the electroslag welding process is shown in Fig. 3-21. Molding shoes are used to form a cavity with the parts to be welded; this cavity contains the molten flux pool, the molten weld metal, and the solidified weld metal. Shielding from the atmosphere is provided by the pool of molten flux. The consumable guide variation is the simplified version of electroslag welding. This variation is shown in detail in Fig. 3-22. The electrode is directed to the bottom of the joint by a guide tube. The guide carries

*®Hobart Brothers Co.

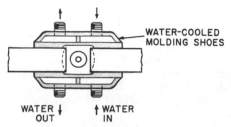

Figure 3-21 Process diagram of electroslag welding (top view).

the welding current and transfers it to the electrode which passes through its hollow core. The guide tube melts off just above the flux bath. The electrode wire protrudes into the molten flux bath and gradually melts as it is fed deeper into the molten pool. The melted metal from the guide tube, from the electrode wire, and from the edges of of the joint collect at the bottom of the flux pool and form the molten weld metal. The molten weld metal slowly solidifies and joins the parts being welded. There is no arc except at the start of the weld before the granulated flux melts from the heat of the arc to become the molten slag. Welding is done with the axis of the joint in the vertical position. The molding shoes are usually water-cooled, and their surface determines the contour of the finished weld. The shoes are fixed and nonsliding. The other version of ESW, not using the con-

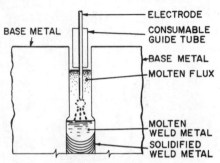

Figure 3-22 Process diagram of electroslag welding (side view: consumable guide tube).

sumable guide, uses sliding shoes that move upward along the joint as the weld is made. The electrode feed head is usually mounted above the weld joint and does not move. For welding thicker materials, the head may be oscillated to provide a wider joint. Extra electrodes and guides may be employed for making extra-wide joints. The square-groove weld joint is normally used. The welding process is limited to a minimum thickness of ¾ in (20 mm) and a maximum thickness with one power source of 3 in (7.5 cm). With two wires and additional power sources, the thickness can be increased to 12 in (30 cm). Once the system is started it continues automatically until the weld is completed. Additional flux is added until the joint is completed. Starting tabs and runoff tabs are employed. They are removed when the weld is completed. A thin slag covering, which is easily removed, adheres to the surface of the welds.

Application

The electroslag process using the consumable guide version has the following features: (1) extremely high metal deposition rates, (2) ability to weld thick materials in one pass, (3) joint preparation and fit-up requirements more tolerant than those for other arc-welding processes, (4) little or no angular distortion, (5) electrode utilization approaching 100 percent, and (6) flux consumption much lower than with the submerged arc process. In addition, once the weld is started, the process is continuous until the weld is completed.

The electroslag process with the consumable guide will weld low-carbon steels, low-alloy high-strength steels, medium-carbon steels, alloy steels, and stainless steels. Quenched and tempered steels can be welded; however, subsequent heat treating is required to maintain weld-joint properties. This is because of the slow cooling cycle inherent in the process. The process can be used for welding joints from as short as 4 in (10 cm) to as long or as high as 12 ft (4 m). A single electrode and guide are used on materials ranging from ¾ to 2 in (20 mm to 5 cm) in thickness. For materials from 2 to 5 in (5 to 12 cm) in thickness, the electrode and guide tube are oscillated in the joint. With material from 5 to 12 in (12 to 30 cm), two electrodes and guide tubes are used and are oscillated in the joint. Oscillation ensures an even distribution of heat in the joint and maintains uniform penetration into the base metal. Use of oscillation reduces the number of electrodes required for welding thicker materials.

Equipment

The major components required for the consumable guide version of electroslag welding are shown in Fig. 3-23. These are (1) the welding machine or power source (one required

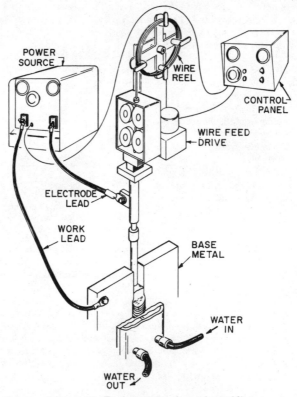

Figure 3-23 Equipment for electroslag welding.

per electrode), (2) the electrode feed head or wire feeder and control system, (3) the mounting device and oscillation mechanism, when required, and (4) the molten-metal-retaining shoes, usually water-cooled. The consumable guide version allows the welding equipment to be taken to the work and is normally mounted on the work itself.

Welding Machine.   The welding machine or power source is normally a CV dc transformer-rectifier machine. The power source must be rated at 100 percent duty cycle and must include a contactor and provisions for remote adjustment. Power sources ranging in size from 500 to 1000 A, CV, are used. DCEP is normally employed. Some variations of the process use ac power sources. The CV-type power source used for electroslag welding can be used for SAW, GMAW, and FCAW.

Electroslag Flux. The flux used for ESW must be designed specifically for the process. It must have a balanced composition to provide the proper electric conductivity in the molten state as well as the proper viscosity and melting temperature. It should have satisfactory shielding characteristics and must have specific deoxiding properties. There is no standard specification for electroslag flux. The amount of flux consumed depends on the fit of the molding shoes to the face of the weld joint. It is normally a fairly constant amount and can be determined by the face width and length of the weld.

Electrode Wire. Electrode wire used for ESW must be matched to the base metal being welded. Normally, electrode wires can be of a composition similar to those used for CO_2 gas-shielded GMAW. Electrode wire must have a minimum covering of copper to minimize seizing to the inside of the guide tube. The electrode is normally solid wire, and the most popular size is ³⁄₃₂ in diam. AWS Specification No. A5.25 is used for specifying electrode wires for electroslag welding.

Guide Tube. The consumable guide version of ESW requires a guide to transmit the electrode wire to the bottom of the weld-joint cavity. Normally, a heavy-wall seamless tube can be used. The size is usually ⅝ in OD and the composition must match the composition of the metal being welded. The guide tube melts during the process and contributes a small portion of the deposit welding metal. Guides of other shapes than round tubes can be employed. In some cases the guide tube is coated with a covering which matches the composition of the electroslag flux. When bare tubes are used and if the joint is extremely long, the consumable guide tube may be fitted with intermittent insulating materials to avoid short-circuiting against the side of the joint.

Electrode wire is supplied in coils or large reels or from payoff packs. Appropriate reel-dispensing equipment is required.

Plasma Arc-Welding Process

Process

Plasma arc welding (PAW), sometimes referred to as needle arc or microplasma, is an electric arc-welding process which fuses the parts to be welded by heating them with a constricted arc between the electrode and the work (transferred-arc mode). Shielding is obtained from the hot ionized gases issuing from the torch orifice. Auxiliary inert shielding gas or a mixture of inert gases supplements the shielding gas system. Pressure may or may not be used, and filler metal may or may not be used. *Plasma*, the fourth state of matter, is defined as a gas which has been heated to a high enough temperature to become ionized. When it is ionized, the gas, or plasma, becomes electrically conductive.

The process, shown in Fig. 3-24, is commonly applied manually but may be applied as a machine or fully automatic process. It can be used to weld almost all metals and can be used in all positions at lower currents. It is normally used on thinner materials. The process requires a fairly high degree of welder skill for manual application and a knowledge of the equipment. There are two ways of using the PAW process. One is known as the melt-in technique and the other, the keyhole technique. The melt-in technique is very similar to gas tungsten arc welding. The keyhole technique actually makes a hole which is then filled as welding progresses. Process details in the keyhole mode are shown in Fig. 3-25.

Application

PAW is similar to GTAW. There is one major difference in that the arc in PAW between the electrode and the work is constricted and forced to go through a small hole or orifice in the torch. A gas is also forced through the orifice; this creates the plasma. The temperature of the plasma is considerably higher than the temperature of the gas tungsten

Figure 3-24   Application of plasma arc welding. (*Hobart Brothers Company.*)

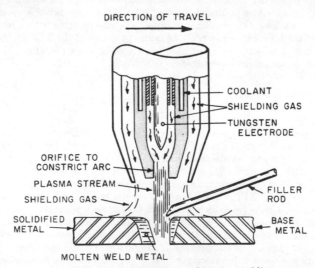

Figure 3-25   Process diagram for plasma arc welding.

arc. The process diagram (Fig. 3-25), shows the tungsten electrode inside the torch and the plasma extending through the torch orifice to the work. The plasma fed through the orifice is ionized and has a columnar form rather than the flare common with GTAW. The ionized gas, or plasma, travels at extremely high speeds and has a force action on the base material, which is important when using the keyhole technique.

One of the advantages of PAW over GTAW is the columnar structure of the plasma which reduces the effect of changes in torch-to-work distance. The high-velocity, high-temperature plasma causes deep penetration in the base metal and allows full penetration of keyhole, single-pass, butt-welding joints. The welds produced have unusually deep penetration with a relatively narrow bead width. The plasma process will weld all the metals that are welded with the GTAW process.

Equipment

The major components required for PAW are shown in Fig. 3-26 and include (1) a welding machine or power source, (2) a special plasma arc console which contains the control

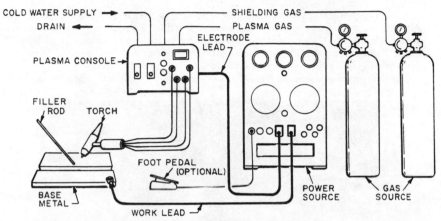

Figure 3-26   Equipment for plasma arc welding.

system, (3) the plasma welding torch, (4) the source of plasma and shielding gas, and (5) filler material when required.

Welding Machine.   The power source for PAW is the CC type with a drooping output characteristic. A GTAW power source is normally used for plasma welding since it includes a contactor, remote current control, and provisions for shielding gas and cooling water. For more complex weldments, programmed current control including upslope and downslope and pulsing is sometimes used.

Plasma Console.   This unit contains a high-frequency arc starter, a nontransferred pilot-arc current supply, torch protection devices, flow meters, and other meters. It may also include the water control system and protective interlocks to protect the plasma arc torch.

Plasma Torch.   The torch contains a tungsten electrode (usually 2 percent thoriated type) and a nozzle having a constricting orifice. Since the arc is enclosed within the torch, all plasma torches are water-cooled. The torches are either manual or machine type with the smaller sizes restricted to the manual applications. Different sizes are available depending upon the current level to be employed.

Shielding Gas–Plasma Gas.   Inert gas, often argon, is normally used as the plasma gas. In addition, argon, helium, or a mixture of the two is used as an auxiliary gas to shield the arc and arc area from the atmosphere. Argon is more commonly used because it is less expensive and easier to obtain. It also provides for better shielding because it is heavier than air.

Filler Metal.   Filler metal may or may not be used. It is not used on very thin metals but is normally used for normal sheet-metal thickness and heavier material. The composition of the filler metal should match that of the base metal. Welding procedure charts will show the recommended filler material used for different base metals. The size of the filler metal (filler rod diameter) depends on the thickness of the base metal and the welding current. Filler metal is usually added to the puddle manually, but automatic feed can be used.

Stud Welding Process

Process

Stud welding (SW), also known as stud arc welding, is a special-purpose arc-welding process used to attach studs to base metal. Partial shielding is obtained by a ceramic ferrule surrounding the stud. It is a machine-welding process using a special gun that holds the stud and makes the weld. The process is normally used on steels in the flat and horizontal positions. A relatively low degree of welding skill is required.

The SW process, shown in Fig. 3-27, was developed in the mid-1930s to satisfy the

Figure 3-27   Application of stud welding. (*Ohio Edison Co.*)

need to weld brackets or retainers to steel plate, particularly in the shipbuilding industry. It became popular for securing wood decking to steel plate, for attaching bracket hangers, etc.

The operation of the SW process is shown in Fig. 3-28. This operation is as follows: (1) A stud gun holds the stud in contact with the workpiece, and the welding operator presses the gun trigger which causes the welding current to flow through the circuit to the stud, which is an electrode, to the work surface. (2) The welding current activates the solenoid within the gun which draws the stud away from the work surface and estab-

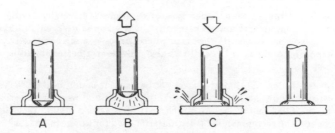

Figure 3-28   Process diagram for stud welding.

lishes an arc. The intense heat of the arc melts the surface of the workpiece and the end of the stud. The length of the arc time is controlled by a timer built into the control unit. (3) When the welding current is automatically shut off, the gun solenoid releases its pull on the stud and a spring action plunges the stud into the molten pool of the workpiece. (4) The molten stud end and the molten pool on the work surface solidify, and the stud weld is completed. The ferrule is broken off and discarded. The process is either a machine application or an automatic application, with machine application being the most popular. For automatic application, the studs are fed automatically into the gun. Welding can be done in all positions; however, flat and horizontal positions are those most commonly used.

Application

SW is widely used for attaching studs and other similar devices to plate or structural members. Studs are normally thought of as threaded, round fasteners; however, rectangular devices, hooks, pins, brackets, and other configurations can be stud-welded to appropriate backing materials. A popular application is the attachment of shear connectors in structural steelwork. Shear connectors are round, usually with a head attached to the upper flange of beams over which concrete is poured, most commonly for bridges and decking. The shear connectors ensure that the steel and concrete work together as a structural member. Another popular application is the use of studs for attaching wood decking over steel decking on ships, particularly aircraft carriers. SW is used for welding connectors used to attach pipe hangers, electric boxes, and other miscellaneous items in ship construction. It is also widely used for attaching insulation to the inside of steel structures. Other uses include attaching and holding insulation to pipe surfaces, attaching studs to hold inspection plates, etc. It is normally used for steels and stainless steels, but variations of the process can be used on nonferrous metals.

Equipment

The major components for SW are shown in Fig. 3-29. These include (1) the welding machine or power source, (2) the stud gun, (3) the control unit, (4) the studs, and (5) the disposable ferrules.

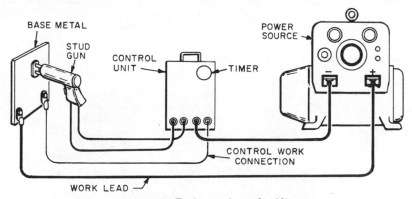

Figure 3-29   Equipment for stud welding.

Welding Machine.   The welding machine or power source is a dc power source which can be a transformer-rectifier or a generator, either motor- or engine-driven. The welding current is dictated by the size or diameter of the stud. The electrode, or the stud, is the negative pole (straight polarity). Amperage required for smaller studs in the $\frac{5}{16}$-in-diam size ranges from 200 to 500 A. With larger size studs, the amperage can be as high as 2300 A. For high current requirements, two or more power sources are connected in parallel. Welding machines for SW should have high overload capacity and a relatively high open-circuit voltage of 95 to 100 V. A CC or drooping-characteristic-type power source is required.

Stud Gun. The stud gun holds the stud and has a switch which starts the control sequence. It also includes the solenoid which provides the withdrawal or lift action to establish the arc. A spring mechanism within the gun applies the pressure required to plunge or push the stud into the pool of the workpiece. The gun should be properly adjusted to accommodate the size of the stud that is being used and to provide the correct arc length during the arc period. The stud gun is normally hand-held and must be held perpendicular to the work. The process can be automated. The stud gun must match, or be of the same make as, the control unit.

Control Unit. The control unit consists of a welding-current contactor, a timing device, and the necessary interconnections. Some control units regulate the speed at which the stud is pushed into the molten base metal; this kind of regulation tends to eliminate spatter and provide more control over the weld shape and quality. The control unit must be of the same type or make as the stud gun.

The welding current passes through the stud gun to provide power for the solenoid.

Studs. Steel studs range in diameter from ⅛ to 1 in (3 to 25 mm) and vary in length; they can be threaded or plain. There are many other types of devices that can be welded with the stud system. These are shown in Fig. 3-30. Studs produced by different manufacturers contain somewhat different fluxing devices on the end of the stud. In most cases, the arcing end contains a charge of welding flux or some other device for shielding the arc area. The fluxes protect the weld and the arc from atmospheric contamination and contain scavengers which purify the melted metal. Essential welding-cycle data are shown in Table 3-1.

Ferrules. A ferrule is used with each stud. Ferrules are made of a ceramic material and are broken off and discarded after each weld is made. They shield the arc area, protect the welding operator, and eliminate the need for a helmet. The ferrule concentrates the heat during welding and confines the molten metal to the weld area. It helps prevent oxidation of the molten metal during the arcing cycle, but it must be made to fit the studs being used. There is no common specification for studs or ferrules, which are manufactured by stud-welding companies.

Air–Carbon Arc Cutting and Gouging Process

Process

The air–carbon arc (AAC) cutting and gouging process is also known as carbon-arc gouging. It is an arc-cutting process in which metals to be cut are melted by the heat of a carbon arc and the molten metal is removed by a blast of air. It is shown in Fig. 3-31. A high-velocity air jet traveling parallel to the electrode hits the molten puddle just behind the arc and blows the molten metal out of the puddle. It is usually a manually controlled operation and can be used in all positions. It can also be applied automatically. The process normally creates considerable noise, and ear protection is recommended. A special electrode holder includes the air-jet opening. The other features of the process are similar to carbon-arc welding. Figure 3-32 shows the details of the process.

Application

The AAC cutting and gouging process is used to cut metal, gouge out defective metal sections, remove old or inferior welds, back-gouge roots of welds, and to prepare grooves for welding. The AAC cutting process is used where slightly ragged edges are not objectionable. It is normally used on steels but can be used on other metals. It is popular for preparing scrap metal for remelting. The surface of some metals deteriorates when cut or gouged by this process. The area of the cut is relatively small since the molten metal

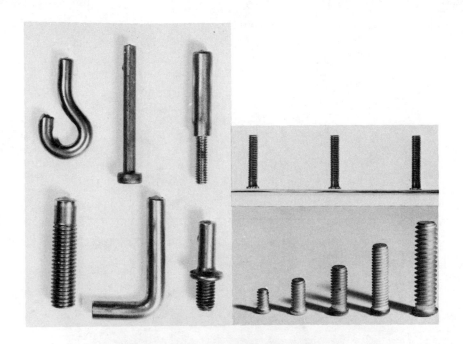

Figure 3-30 Uses for stud welding. (*Nelson Stud Welding Company.*)

TABLE 3-1 Stud-Welding Data

Stud diam, in	Current, A DCEP	Welding specifications			
		Voltage, V	Time, s	Lift, in	Plunge, in
$\frac{3}{16}$	300	30	7	$\frac{1}{16}$	$\frac{1}{8}$
$\frac{1}{4}$	400	30	10	$\frac{1}{16}$	$\frac{1}{8}$
$\frac{5}{16}$	500	30	15	$\frac{1}{16}$	$\frac{1}{8}$
$\frac{3}{8}$	600	28	20	$\frac{1}{16}$	$\frac{1}{8}$
$\frac{7}{16}$	700	28	25	$\frac{1}{16}$	$\frac{1}{8}$
$\frac{1}{2}$	900	28	30	$\frac{3}{32}$	$\frac{5}{32}$
$\frac{5}{8}$	1150	28	40	$\frac{3}{32}$	$\frac{5}{32}$
$\frac{3}{4}$	1600	26	50	$\frac{1}{8}$	$\frac{3}{16}$
$\frac{7}{8}$	1800	24	60	$\frac{1}{8}$	$\frac{3}{16}$
1	2000	24	70	$\frac{1}{8}$	$\frac{3}{16}$

Figure 3-31 Application of air–carbon arc cutting. (*Hobart Brothers Company.*)

is quickly removed. The surrounding area does not reach high temperatures, thus reducing the tendency toward warpage and cracking. In some cases the surface must be ground to provide quality weld-joint preparation.

Equipment

Equipment for cutting and gouging is the same as for carbon-arc welding and for SMAW, with the exception of the special electrode holder and the required compressed-air supply. The necessary equipment is shown in Fig. 3-33. This consists of (1) the welding machine or power source, (2) the special electrode holder or torch, (3) a carbon electrode, and (4) the compressed-air supply.

Welding Machine. The welding machine or power source is normally a CC drooping-characteristic type either a transformer-rectifier or generator. CV machines with flat characteristics may be used, but precautions must be taken to operate them within their

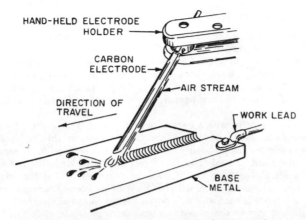

Figure 3-32 Process diagram for air–carbon arc cutting.

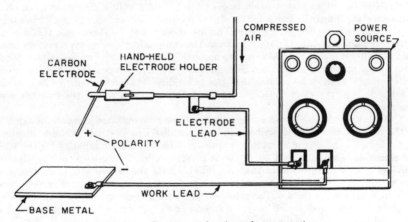

Figure 3-33 Equipment for air–carbon arc cutting.

rated output. Specially designed heavy-duty machines are used for AAC cutting or gouging with large electrodes. Machines of this type are available with a capacity up to 1000 A. The ac power source CC types can be used for special applications; however, ac-type carbons are required.

Electrode Holder. The electrode holder or torch is of a special design which includes the air-jet's stream nozzle and valve. In addition, it must clamp the carbon electrode tightly. Electrode holders come in several sizes depending on the size of the carbon electrode to be used. Larger holders may be water-cooled. The cable assembly includes the compressed-air hose which is connected to the supply.

Electrodes. Electrodes used for AAC cutting and gouging can be of pure carbon or the graphite type. Electrodes are also available with a copper coating which tends to make the electrodes last longer. Electrodes erode away rapidly during heavy-duty cutting. Electrode diameters may be ³⁄₁₆, ¼, ⁵⁄₁₆, ⅜, ½, or ⅝ in. Larger sizes are also available. AWS Specification A5 provides specifications for electrodes for this process.

Air Supply. A supply of dry compressed air is required. The air pressure is not critical and ranges from 80 to 100 lb/in². It is normally obtained from shop lines or from an air compressor.

Other Arc Cutting Processes

The electric arc, a highly concentrated energy source, can be useful for cutting metals. The arc alone does not produce good-quality cuts, but, when assisted by a jet of oxygen or air, or by plasma, the quality of the cut is greatly improved. There are several arc-welding processes which can also be used for cutting. They are the plasma arc, the carbon arc, the arc between covered electrodes and the work, and a system that uses a special, hollow, covered electrode whereby oxygen can be introduced inside the electrode to produce a quality cut.

The principle of cutting with an arc is to melt metal. This is done by increasing the heat input faster than heat is extracted from the arc area. On thin materials, the molten metal will fall away by gravity, and a rather crude cut will result. In emergency situations, the arc alone can be used. This is not recommended for industrial applications.

For arc cutting the parts to be cut must be arranged so that the area beneath the parts will receive the molten metal without creating problems. For carbon-arc cutting, a single carbon electrode is used with a dc power source using the electrode on the negative pole. The carbon should be sharpened to a long taper approximately half its diameter at the end. It should be gripped close to the arc end to avoid overheating. Position the material to be cut so that it projects over a table edge with a container to catch the molten metal. The arc is struck on the edge of the plate, with a fairly long arc maintained until a puddle is melted at the edge. The arc should then be shortened; this will help force the molten puddle to fall away from the material. A sawing-type motion can be used to help remove the metal from the material being cut. The "icicles" which tend to form on the bottom of the cut can be removed by the arc. Holes may be pierced in steel plates up to ⅜-in thick by striking the arc and holding it in one spot until a large puddle is formed. Feed the electrode downward and force it through the plate as the metal becomes molten.

Cutting with the shielded-metal electrode can be accomplished in much the same way as the carbon-arc cutting described above. A smaller-size electrode is used, usually ⅛-, ⁵⁄₃₂-, or ³⁄₁₆-in diam, but the welding machine must have sufficient capacity for the size of the electrode selected. The E 6011 type is recommended. The coating on the electrode provides a little more arc force than is available with the carbon arc. If electrodes are quickly dipped in water prior to use for cutting, they tend to provide more of a cut before they are consumed in the arc. The shielded-metal arc and carbon-arc cuts are extremely rough and are normally used only for emergency situations when the proper cutting equipment may not be available.

Oxy arc cutting is a proprietary method using covered electrodes having a hollow core. A special electrode holder is required. This electrode allows oxygen to pass through the hole in the center of the electrode. A valve on the electrode holder allows the oxygen to be started or stopped. In this process, the arc is struck in the normal way and, as soon as the metal becomes molten, the oxygen is turned on, this provides a jet that oxidizes the molten metal and carries it away. This process is extremely useful for cutting materials such as cast iron, high-chromium stainless steels, and other hard-to-cut substances.

GAS WELDING AND CUTTING

Oxyacetylene Gas Welding Process

The oxyacetylene gas welding (OAW) process—sometimes called gas welding, oxyfuel gas welding, or torch welding—is an oxy-fuel gas process used to fuse the parts to be welded by heating with a gas flame or flames obtained from the combustion of the acetylene with oxygen. The process, shown in Fig. 3-34, may be used with or without filler metal added to fill gaps or grooves. It can be used on thin- to medium-thickness metals of many types. It is most commonly used on nonferrous metals and can be used in all positions. It is applied as a manual process and requires a relatively high degree of skill on the part of the welder.

Figure 3-34 Application of oxyacetylene welding. (*Hobart Brothers Company.*)

Oxyacetylene welding is the oldest of the modern welding processes. It came into popularity in the late 1800s and is used for repair and maintenance, overlaying, sheet-metal welding, and small-diameter-pipe welding.

The oxyacetylene welding process is diagrammed in Fig. 3-35. The oxyacetylene flame

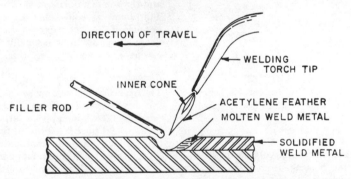

Figure 3-35 Process diagram for oxyacetylene welding.

is extremely hot, approaching 6300°F (3200°C). This hot flame melts the surface of the materials to be joined so that they flow together to produce a weld. Filler material in the form of a rod is added to fill gaps or grooves. The mixing of the oxygen and fuel gas takes place in the welding torch, and the flame is initiated by means of a spark lighter. An atmosphere provided by the burning of the gases shields the molten metal from the atmosphere.

Process

The oxygen and acetylene flow through the hoses from the supply source or from individual cylinders to the welding torch, where they are mixed and burned at the torch tip.

The reaction for this is

$$2C_2H_2 + O_2 \rightarrow 4CO + 2H_2$$

This is the primary reaction that occurs in the inner cone of the flame adjacent to the welding tip. The secondary reactions, as shown,

$$4CO + 2O_2 \rightarrow 4CO_2$$
$$2H_2 + O_2 \rightarrow 2H_2O$$

occur in the outer portion of the flame, and the extra oxygen is obtained from the atmosphere. Note that CO_2 and water vapor result from the secondary reactions. The CO_2 formed shields the molten metal from the atmosphere.

In the combustion of oxygen and acetylene the gases are mixed in the torch in about equal volume, but the remaining oxygen required is from the atmosphere. When the proportions of oxygen and acetylene from the cylinders are the same, the type of flame is referred to as a neutral flame (shown in Fig. 3-36). The exact adjustment is such that the inner cone is just defined, with no feather of acetylene appearing in the flame. This type of flame is used mostly for welding, brazing, and heating. When slightly more acetylene is applied to the neutral flame, a visible feather is seen extending from the inner cone, as shown in Fig. 3-36. This is known as a reducing flame. It has excess acetylene and is used for welding alloy steels, aluminum, and cast iron or for certain surfacing applications. It is slightly cooler than the neutral flame.

When additional oxygen is supplied, the inner cone becomes darker and shorter and the entire flame is smaller and hotter. This flame is called an oxidizing flame because of the excess oxygen. An oxidizing flame (Fig. 3-36) is normally not used except for oxygen flame cutting.

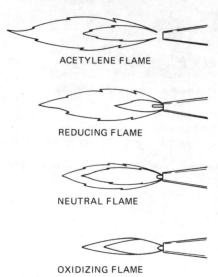

ACETYLENE FLAME

REDUCING FLAME

NEUTRAL FLAME

OXIDIZING FLAME

Figure 3-36   Flame types for oxyacetylene welding.

Application

The oxyacetylene welding process has certain advantages: (1) the equipment is very portable, (2) it is a highly versatile process, (3) the weld pool is visible to the welder, (4) welding is possible in all positions, (5) the equipment is relatively inexpensive, and (6) the same basic equipment can be used for welding, heating, torch brazing, and oxygen flame cutting. The main disadvantage of oxyacetylene welding is the fact it is relatively slow and relatively expensive to use because of the prices of the gases. Oxyacetylene welding is most useful for joining thin (up to ¼-in) steel, copper, and copper alloys, and it can be used for welding aluminum and other nonferrous alloys. It can also be used for overlaying, for surfacing, for wear resistance, etc. as well as for heating metals for bending, straightening, etc. Its industrial applications include maintenance and repair, auto-body repair, welding small-diameter piping, brazing, and light manufacturing.

Equipment

Equipment for oxyacetylene welding includes: (1) the welding torch and tips, (2) the hose for transporting the gas from the supply to the torch, (3) regulators for oxygen and acetylene (normally attached to cylinders or to the supply-pipe system), (4) a cylinder or

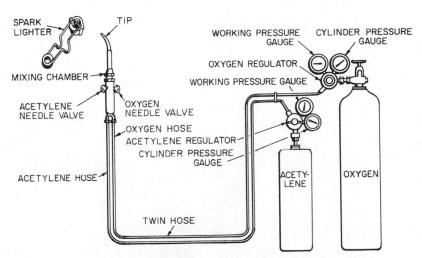

Figure 3-37   Equipment for oxyacetylene welding.

supply of oxygen, and (5) a cylinder or supply of acetylene. In addition, a spark lighter, torch, and cylinder wrench are required. Often, a cylinder cart for transporting the cylinders and apparatus is used. This arrangement is shown in Fig. 3-37.

Welding Torch.   The welding torch, sometimes called a *blowpipe,* is the major piece of equipment for the process. It performs the function of mixing fuel gas with oxygen to produce the required type of flame, which is then directed manually as desired. The torch consists of a handle, or body, which contains the hose connections for oxygen and acetylene. It also contains the oxygen and acetylene valves (sometimes called *needle valves*) for regulating gas flow into the torch and a mixing chamber. A medium-pressure oxyacetylene torch is shown in Fig. 3-38. Different sizes of tips can be attached to the torch.

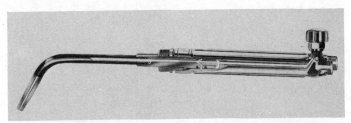

Figure 3-38   Medium-pressure torch for oxyacetylene welding. (*Smith Torch Company.*)

These are identified by manufacturers' numbers which indicate the hole or orifice in the end of the tip. Unfortunately, there is no standard system for identifying tip sizes, and each manufacturer has its own system; however, in every case the system relates to a drill size which identifies the diameter of the hole.

Gases.   The gases used for oxyacetylene welding are oxygen and acetylene. Acetylene produces the highest-temperature flame and is considered the all-purpose fuel for this process. Acetylene is colorless but it has an easily detected odor somewhat like the odor of onions. When the torch is used for heating, other fuel gases may be used, such as natural gas, propane, and proprietary fuel gases. Different tips and mixers are usually needed when other gases are used.

Regulators. The pressure of the gas used for oxyacetylene welding is relatively low; however, the pressure of the gas in the supply system or in individual cylinders is relatively high. Therefore, a device known as a *gas regulator* is used to reduce the pressure from the high side to the correct working pressure for the torch. This is a complex unit, made of needle valves, springs, and diaphragms, for precisely producing the lower pressure used by the torch. Figure 3-39 shows a gas regulator. The regulators for oxygen and

Figure 3-39 Gas regulator for oxyacetylene welding. (*Hobart Brothers Company.*)

acetylene are different and *cannot be interchanged.* Oxygen connections have right-hand threads and acetylene and other fuel gas connections have left-hand threads. A gas regulator will keep the gas pressure constant and has gauges showing the pressure going to the torch and sometimes showing the pressure of the supply. Two-stage regulators are normally used with cylinders, and single-stage regulators are normally used for supply lines.

Gas Cylinders. Oxygen and acetylene are both supplied in individual cylinders. A pair of cylinders used for oxyacetylene welding is shown in Fig. 3-40. Oxygen cylinders are made of a high-strength steel and contain oxygen at a very high pressure: up to 24,000 lb/in^2. *CAUTION—Cylinders must be treated carefully and inspected periodically. Mistreating cylinders can damage them and may cause them to explode, creating a very dangerous situation.*

Acetylene, on the other hand, is stored at a relatively low pressure. Acetylene cannot be stored safely over 15 lb/in^2. It is dissolved in liquid acetone which is contained by a filler material in the cylinder. An acetylene cylinder will have a working pressure of 250 lb/in^2; however, most of the acetylene is dissolved in the acetone, which keeps it stable and eliminates the danger from high-pressure free acetylene. Acetylene cylinders should always be kept upright because of the liquid inside the cylinder. In addition, an acetylene cylinder should always be kept away from high temperatures and should be treated with the respect due *any* gas cylinder. There is no uniform national color code for gas cylinders. Each company supplying gases has its own color code. However, there is standardization of threads on the fittings of the cylinders; remember, oxygen cylinders have right-hand threads and acetylene cylinders have left-hand threads.

Cylinder Carts. For portable installations, cylinder carts are usually employed. This allows the cylinders to be affixed to a structure even though it is portable. It allows the storage of the hoses and torch and is useful for maintenance applications.

Safety Precautions

The safety precautions for oxyacetylene and gas welding are somewhat special for the process. For your own safety and the safety of those about you, it is important to follow these safety directions when you are using oxyacetylene welding equipment.

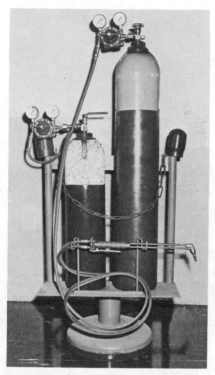

Figure 3-40 Gas cylinders for oxyacetylene welding. (*Hobart Brothers Company.*)

Figure 3-41 Application of oxy–fuel gas cutting. (*Hobart Brothers Company.*)

For additional information concerning the oxyacetylene welding process refer to Sec. 6-2 of Howard Cary's *Modern Welding Technology,* Prentice-Hall, Englewood Cliffs, N.J., 1979.

Oxy–Fuel Gas Cutting Process

Process

The oxy–fuel gas cutting (OFC) process, also known as oxygen cutting, gas cutting, burning, and so on, is a thermal process used to sever metals by heating the metal with a flame to an elevated temperature and then using pure oxygen to oxidize the metal and produce the cut. Different fuel gases can be used, including acetylene, natural gas, propane, and a variety of proprietary or trade-name fuel gases. The process shown in Fig. 3-41 can be applied manually or by machine. It can be used to cut ferrous materials in sections varying from thin to thick, and it can be used in all positions. Manual OFC requires a fairly high degree of skill to produce quality cuts.

Details of the process are diagrammed in Fig. 3-42. This diagram shows the torch and cutting tip, the preheating flames to bring the metal up to the kindling temperature, and the oxygen jet supplied to oxidize, or "burn," metal away to produce the cut.

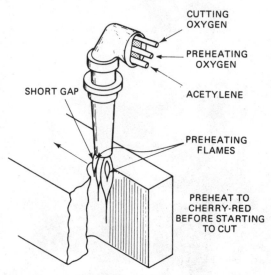

Figure 3-42   Process diagram for oxy–fuel gas cutting.

Application

This cutting process (1) is very portable, (2) is versatile, (3) allows cutting in all positions, (4) uses relatively inexpensive equipment, and (5) can be used to cut steels. The disadvantages of the process are: (1) it cannot be used to cut nonferrous materials, and (2) the cut surfaces are not as smooth as mechanically cut surfaces. It is widely used throughout industry as a manual process. It is also widely used as a machine-cutting process with automatic torch controls and seam-following devices. When it is used as an automatic process, extremely smooth surfaces can be obtained.

Equipment

Oxy–fuel gas welding equipment includes: (1) the cutting torch and tips, (2) oxygen and fuel gas hoses, (3) regulators for oxygen and fuel gas or acetylene, and (4) a supply of oxygen and fuel gas from cylinders or a piping system. The equipment is essentially the same as used for oxyacetylene welding.

The Cutting Torch.   The cutting torch can be a combination cutting and welding torch or a torch especially designed for cutting only. The gases are mixed within the torch, and needle valves control the quantity of each gas flowing into the mixing chamber. A lever-type valve controls the oxygen flow for cutting. Various sizes and types of tips are used with the cutting torch for specific applications of cutting, gouging, beveling, etc. The cutting tips are sized by the oxygen orifice size in the cutting tip. There is no standard cutting-size designation, and each company has its own system; however, each cutting tip size relates to the standard drill size for the cutting orifice. In this way they can be related to the thickness of metal to be cut. Preheat flames are arranged around the central cutting orifice and are sufficient to bring the metal to the kindling temperature prior to cutting.

The rest of the equipment is the same as that used in the oxyacetylene welding process.

Gases.   The gas used for oxygen cutting is normally pure oxygen, while the fuel gas is always a hydrocarbon gas, often acetylene. Other fuel gases used are natural gas, propane, and a variety of proprietary liquid-petroleum-base or propane-base gases. The selection of fuel gas is extremely complex; however, the fuel gas is used for the preheating flame that brings the material to be cut up to its kindling temperature. The basic cutting

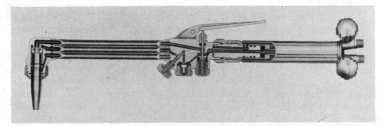

Figure 3-43 Cutting assembly with welding torch body. (*Smith Torch Company.*)

process using oxygen is not affected by the choice of preheat fuel gas that is used. The type of fuel gas relates primarily to the time period required to bring the material up to the kindling temperature. Figure 3-43 shows the flame-cutting assembly.

Safety Precautions

Safety precautions for oxygen–fuel gas cutting are extremely important because it is widely used in maintenance and construction work. The normal precautions involving gases under pressure should be observed. Additional precautions relate to the cutting of vessels and containers that may be sealed and/or may have contained combustible materials. This should not be done without taking extra-special precautions. Another problem relating to oxygen cutting is the fact that white-hot metal from the cut will travel many, many feet and will retain sufficient heat to set combustible materials on fire. Metallic or noncombustible material should be used to backstop the hot metal being ejected from the cut. Cutting should *never* be attempted in confined areas without first testing the atmosphere and providing a fire watch or observers to continually watch the cutter while it is in operation.

BRAZING AND SOLDERING

Distinction

The primary difference between brazing and soldering is the arbitrary temperature of 840°F (450°C). Both are a group of processes which join materials by heating them to a suitable temperature; both use a filler metal which is distributed between closely fitted surfaces of the joint by capillary attraction. Solder, the filler metal used in soldering, has a melting temperature below 840°F (450°C), while brazing alloy, the filler metal used for brazing, has a melting temperature above 840°F (450°C). Both solder and brazing alloy have a composition somewhat different from that of the base metal. Also, for both soldering and brazing, a fluxing material is normally used.

There is one other term that should be mentioned—*braze welding*. It refers to a process that is similar to, but different from, brazing in that capillary attraction is not used to distribute the filler metal. Braze welding is used quite often to join cast-iron sections.

Heating

The method of heating the materials to be joined is the method usually used to differentiate between the different soldering and brazing processes. The same methods of heating can be used for both soldering and brazing. These include using a gas torch, which is one of the most common methods of both brazing and soldering; dipping the materials in flux or molten filler metal; generating heat in the parts by means of a furnace; applying heat by means of induction or infrared radiation; generating heat by means of the resistance of the parts to current flow; and introducing heat by means of an iron—a method used only for soldering.

Torch brazing is the method discussed here for applying heat to the parts to be joined. In torch brazing, heat can be applied by using different fuel gases and oxygen or air combinations.

The torch for melting high-temperature brazing alloys is the same as that used for oxy–fuel gas welding, whereas the torch for soldering uses a fuel-gas–air system. Different torches are used for the different fuel gas and oxygen or air combinations. In each case, however, the use of the torch and its manipulation are essentially the same. The basic principle is to provide uniform heating of the parts being joined. Proper fluxing and proper fit of the parts are essential to allow capillary attraction to pull the molten filler metal into the joint.

Joints

Both brazed and soldered joints require close fit of the parts to be joined. This is necessary to provide the capillary attraction to pull the alloy filler metal into the joint to provide sufficient area of filler metal to ensure a sufficiently strong joint. The lap-type joint is most commonly used since it provides for sufficient faying surfaces to attract the filler material. Butt-type joints are rarely used for soldering or brazing. One of the most common types of joints is the socket joint used for pipe and tubing.

Fluxes

Flux is almost always used in torch brazing. The flux helps maintain cleanliness of the faying surfaces so that the filler metal will adhere properly. The joints should be properly cleaned before applying flux because cleaning the surface *is not* the function of the flux. It does, however, help by combining with, dissolving, or inhibiting the formation of chemical compounds which would interfere with the quality of the joint. The flux also protects the surface during the heating operation. The type of flux to be used is chosen on the basis of the process function and metal to be joined. The flux and the filler metal must also be matched. Manufacturer's information concerning fluxes must be followed since there are no established specifications covering fluxes. However, the American Welding Society provides flux type numbers with recommendations for use. AWS Specification A5.8 on filler metals for brazing and ASTM Specification B 32 on the composition and uses of solders provide further information. The particular alloy or type of filler metal to be used depends upon the process and the metals being joined.

Summary

Quality brazes or soldered joints can be made by following the basic principles of cleanliness, fluxing, joint detail, and matching the proper flux and filler metal alloy.

OTHER WELDING PROCESSES

The previous sections provided information concerning arc-welding processes and some of the other welding and cutting processes most commonly used by plant engineers. However, there are many other welding processes used in manufacturing that should be mentioned.

The American Welding Society's "Master Chart for Welding and Allied Processes" (Fig. 3-44) shows seven families of welding processes, two families of allied processes (thermal spraying and adhesive bonding), and three families of thermal cutting processes. Previously we have described arc welding, brazing, soldering, oxy–fuel gas welding, oxygen cutting, and arc cutting. Some other important processes are now briefly described.

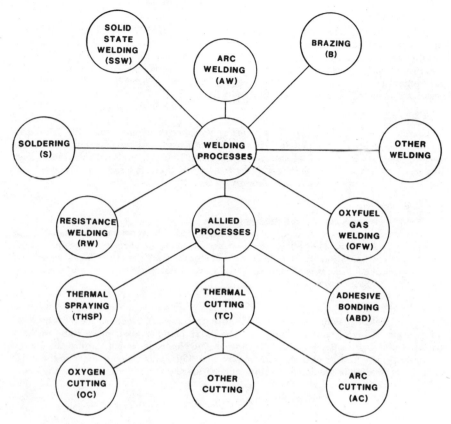

Figure 3-44   American Welding Society master chart of welding and allied processes.

Resistance Welding

Resistance welding is a group of welding processes that produce joints of metal by means of heat obtained from resistance and pressure. The resistance is that of the work to the electric current in a circuit of which the work is a part, and the pressure is applied externally. Spot welding is the most popular of the resistance-welding processes. Spot welding is accomplished with a machine which uses electrodes to carry the current to and through the joint being welded. The electrodes also apply the pressure which is necessary to force the parts together after the current has heated the metal to the welding temperature. Resistance welding is extremely fast, and filler metal is normally not required. It is very popular for welding automobile bodies and for making household appliances. Other resistance-welding processes are projection welding, seam welding, flash welding, and high-frequency resistance welding, with many variations of each. Most metals can be resistance-welded. Special precautions are required, however, for certain metals.

Solid-State Welding

The solid-state family of welding processes includes friction welding, cold welding, ultrasonic welding, and other less-important processes.

Friction Welding

In friction welding, the weld is produced by heat obtained from a mechanical sliding motion between rubbing surfaces. The process usually involves rotating one part against another to generate frictional heat at the junction. When a suitably high temperature has been reached, rotational motion ceases and pressure is applied to create the weld. Equipment is similar to a lathe: it is extremely fast, and no filler metal is required. It is restricted primarily to mass-production industries.

Cold Welding

Cold welding is a solid-state process whereby pressure at room temperature is used to produce the weld. The metals are substantially deformed, and extremely high pressures are required on extremely clean interfacing surfaces. This process is restricted to thinner materials, and it is often used for welding nonferrous materials such as aluminum and copper. It is also used to weld aluminum to copper.

Explosion Welding

Explosion welding is also a solid-state process. In this case the weld is obtained by high-velocity movement together of the parts to be joined. The movement is caused by an explosion, and heat is not applied. The interface between the parts welded shows a saw-tooth-type configuration. Heat is instantly produced from the shock wave associated with impact. This process is often used to weld dissimilar parts together and is used for overlaying or cladding materials.

Ultrasonic Welding

Ultrasonic welding is another of the solid-state processes. It produces the joint by local application of high-frequency energy to the parts being welded, while they are held together under pressure. Welding occurs when the electrode, which couples the energy to the work, is vibrating at ultrasonic frequencies. This, plus pressure, creates the weld. Ultrasonic welding is restricted to thinner materials and is quite often used in the packaging industries.

Electron-Beam Welding

Electron-beam (EB) welding is one of the most important non-arc welding processes. In EB welding, the heat for welding is obtained from a concentrated beam of high-velocity electrons impinging upon the surface of the work. Pressure is not used, but filler metal is sometimes added.

EB welding was initially done in a vacuum chamber. The work and work-moving devices, as well as the electron beam, were contained in the chamber. The electron beam is generated by an electron gun and is similar to that in an x-ray tube. The work had to be taken to the machine and it had to fit within the chamber. Evacuation of the chamber was a major part of the operation. Recently, however, specially designed chambers which allow continuous entrance and exit of parts have been used in mass-production industries. A lower vacuum in the chamber is sometimes used. EB welding in the atmosphere is now also possible. However it is restricted to operating close to the electron gun, which must be in a vacuum chamber. The capital expense for EB welding is quite high, and this type of welding, therefore, is restricted to specialty materials and special applications.

Laser-Beam Welding

Laser-beam welding is very similar to electron-beam welding except that the heat is obtained from the application of a concentrated coherent light beam impinging on the surface of the work; a vacuum chamber is not required. However, the generation of a laser beam is extremely complex and expensive and the electrical efficiency of the process is relatively low. This process is quite new, and additional developments are expected. The laser beam is used for cutting as well as for welding and will cut nonmetals as well as metals. At this time, there are more applications for laser cutting than for laser welding. However, developments in this field are accelerating and the reader is urged to investigate the state of the art when considering the various alternatives for any application.

Thermite Welding

One of the older welding processes still in use is thermite welding. In this process, the weld is produced by heating the parts to be joined with superheated liquid metal obtained from a chemical reaction between a metal oxide and aluminum. The filler metal is obtained from the superheated liquid metal. The heat is obtained from an exothermic reaction between iron oxide and aluminum. This reaction occurs immediately above the weld, and when it has gone to completion, the superheated liquid flows into the weld area and is retained by a mold. The process is used for joining rails, reinforcing bars, and other similar items. It is also used to join castings used in ship construction.

QUALITY CONTROL AND INSPECTION METHODS

The quality of welds can be determined by nondestructive testing methods. Welds made in most commercial metals normally equal the strength of the base metal. This result depends upon the proper selection of the process and procedure, including the filler metal. Welds in metals having special properties resulting from heat treatment or working may not equal the strength of the base metal because the heat of making the weld will cause these special properties to deteriorate in the area adjacent to the weld. For these types of metals special precautions are required. For all other welds, however, the quality of the weld can be determined and controlled. Adherence to procedures that are known to produce quality welds is recommended. After the weld has been made, it can be inspected by a number of nondestructive evaluation techniques. The most popular is visual inspection. Visual inspection (VT) is used by welders, supervisors, and inspectors for potential defects such as undersized welds which can be checked by gauges, rough or irregular surface, surface cracks, surface porosity, undercut, etc. In addition, weld quality can be determined by at least four other evaluation techniques. These are summarized in Table 3-2.

Visual Inspection

Process and Applications
Visual welding inspection is the most widely used and most valuable welding inspection technique. In particular, it is the most effective for noncritical welding production. Visual inspection requires less time than any other inspection method and is also the least expensive. In addition to being a weldment inspection technique, it allows inspection of the welding procedures themselves and thus is also a preventive tool. The inspector is able to watch and require procedure conformity during weldment production.

Visual inspection throughout the forming of a single weldment can catch errors in each step and items which might develop errors, such as faulty materials and procedures. Repairs can be made on an incompleted piece of work. Inspectors can check the basic materials, the joint preparation, process manipulation, and welding technique long before the weldment is completed. This prevention and early correction of the welds is particularly important on highly critical or expensive weldments.

Inspectors can note errors in weld preparation, dimensions, alignment, fit-up, cleanliness, welding procedure, warpage, finish, and mishandling in marking. They can detect scabs, seams, scale, surface slag, laminations, roughness, spatter, craters, surface porosity, undercuts, overlaps, cracks, and inadequate penetration. They can check for many of these at once and can note several defects simultaneously.

For any other welding inspection technique, inspectors need, primarily, to be able to interpret a series of indicators. With visual inspection, they must know welding more

TABLE 3-2 Guide to Welding Quality Control (NDT) Techniques

Technique	Equipment	Defects detected	Advantages	Disadvantages	Other considerations
Visual, VT	Pocket magnifier, welding viewer, flashlight, weld gauge, mirror	Weld preparation, fit-up; cleanliness, roughness, spatter, undercuts, overlaps, inadequate penetration and size; welding procedures	Easy to use; fast, inexpensive, usable at all stages of production	For surface conditions only; dependent on subjective opinion of inspector	Most universally used inspection technique
Magnetic particle, MT	Iron powder, wet, dry, or fluorescent; commercial power source; black light for the fluorescent type	Surface and near-surface discontinuities, cracks, etc.; subsurface porosity and slag on light materials	Indicates discontinuities not visible to the naked eye; useful in checking edges prior to welding; also, repairs; no size restriction	Used on magnetic materials only; surface roughness may distort magnetic field	Testing should be from two perpendicular directions to catch discontinuities which may be parallel to one set of magnetic lines
Liquid penetrant, PT	Fluorescent or visible commercial penetrating liquids and developers; black light for the fluorescent type	Defects open to the surface only	Very small, tight, surface imperfections show up. Easy to apply and to interpret; inexpensive; use on either magnetic or nonmagnetic materials	Somewhat time-consuming in the various steps of the processes	Often used on root pass of highly critical pipe welds; if material improperly cleaned, some indications may be misleading
Radiographic, RT	X-ray or gamma-ray equipment; film-processing equipment; film-viewing equipment; penetrometers	Most internal discontinuities and flaws; limited by direction of discontinuity	Provides permanent record; indicates both surface and internal flaws; applicable on all materials	Usually not suitable for fillet-weld inspection; film exposure and processing critical; slow and expensive	Most popular technique for subsurface inspection; required by many codes and specifications
Ultrasonic, UT	Commercial ultrasonic units and probes; reference and comparison patterns	Can locate all flaws located by other methods with the addition of other exceptionally small flaws	Extremely sensitive; use restricted to only very complex weldments; can be used on all materials	Time-consuming; demands highly developed interpretation skill; permanent record not normally obtained	For irregularly shaped parts, immersion testing often used; required by some codes

thoroughly and be able to inspect all areas of the weldment production. The technique depends upon the alertness, eyesight, welding knowledge, and subjective judgment of the inspectors.

Visual inspection is unreliable on subsurface conditions and discovery of these must result primarily on the inspectors' judgment of the welders' actual work. Tiny, fine flaws can be overlooked very easily and can be covered by peening and hammering while removing slag.

Because of the simplicity, absence of elaborate equipment, and low cost of visual inspection, it can be relied upon too heavily when used entirely by itself rather than in conjunction with more sensitive inspection methods.

Equipment

A pocket magnifier, flashlight, borescope, dentist's mirror, weld gauge, straightedge, T-square, and weld standards are all helpful pieces of equipment in visual inspection.

WELDING CODES AND QUALIFICATIONS OF WELDERS

Before a welder can begin work on any job covered by a welding code or specification he or she must become certified under the code that applies. Many different codes are in use, and it is exceedingly important that the specific code is referred to when one is taking qualifying tests. (Standard welding symbols as shown in Fig. 3-45 are used throughout the industry.) In general, the following types of work are covered by codes: pressure vessels and pressure piping, highway and railway bridges, public building, tanks and containers that will hold flammable or explosive materials, cross-country pipelines, aircraft, ordnance material, ships, and boats. A qualified welding procedure is normally required.

Certification is obtained differently under the various codes. Certification under one code will not necessarily qualify a welder to weld under a different code. In most cases certification for one employer will not allow the welder to work for another employer (except in cases where welders are qualified by an association of employers). Also, if the welder uses a different process or if the welding procedure is altered drastically, recertification is required. In most codes, if the welder is continually employed, welding recertification is not required, providing the work performed meets the quality requirement. An exception is the military aircraft code which requires requalification every 6 months.

Qualification tests may be given by responsible manufacturers or contractors. On pressure vessels, the welding procedure must be qualified before the welders can be qualified. Under some codes this is not necessary. To become qualified, the welder must make specified welds using the selected process, base metal, thickness, electrode type, position, and joint design. Standard test specimens must be made under the observation of a qualified person. In government specifications, a government inspector must witness the making of welding specimens. Specimens must be properly identified and prepared for testing.

The most common test is the guided-bend test. In some cases x-ray examinations, fracture tests, or other tests are employed. Satisfactory completion of test specimens, provided that they meet acceptability standards, qualifies the welder for specific types of welding. The welding that is allowed depends on the particular code. In general, the code indicates the range of thicknesses which may be welded, the positions which may be employed, and the alloys which may be welded.

Qualification of welders is an extremely technical subject and cannot be adequately covered here. The actual code must be obtained and studied prior to taking any test.

The most important codes are:

AWS D1.1, "Structural Welding Code"
"Welding Qualifications," Sec. IX of "ASME Boiler and Pressure Vessel Code"
API #1104, "Standard For Welding Pipelines and Related Facilities"

These codes can be obtained from the sponsoring associations.

AMERICAN WELDING SOCIETY

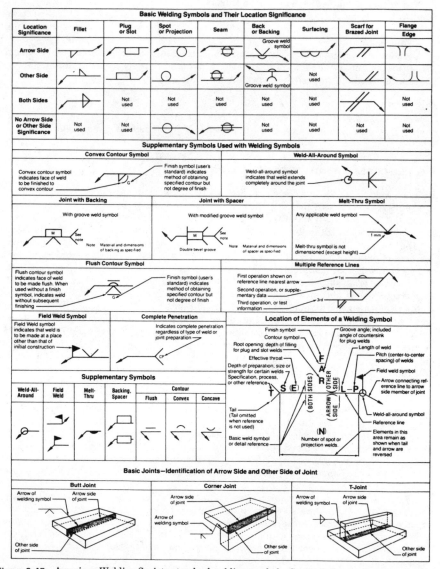

Figure 3-45   American Welding Society standard welding symbols. Continued on pages 1-73 to 1-75.

POWER SOURCES FOR ARC WELDING

Many different types and sizes of arc-welding machines are available. It is important to select the best machine and the one most suited for the particular work to be done. The following information describes the different types of machines available, thus allowing the selection of the one most ideally suited for each type of work.

STANDARD WELDING SYMBOLS

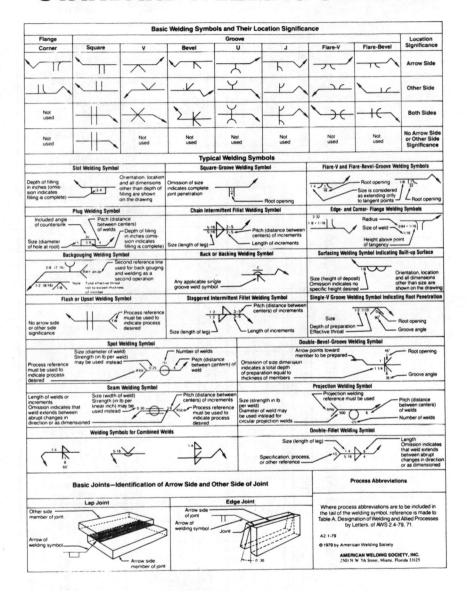

Arc-welding machines can be classified in many different ways, such as rotating machines, static machines, electric-motor–driven machines, internal-combustion-engine–driven machines, transformer-rectifiers, limited-input welding machines, conventional CV voltage welding machines, and single-operator machines or multiple-operator machines.

There are two basic categories of power sources: the conventional or CC or variable-voltage welding machine with the drooping volt-ampere curve and the CV or CP or mod-

AMERICAN WELDING SOCIETY — STANDARD WELDING SYMBOLS

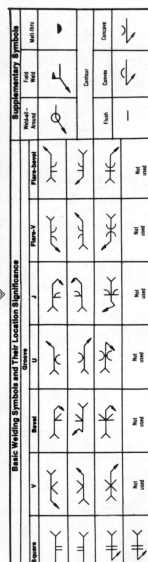

Basic Welding Symbols and Their Location Significance

Square	V	Bevel	Groove		Flare-V	Flare-bevel
			U	J		

Supplementary Symbols

Weld-all-Around	Field Weld	Melt-thru

Contour		
Flush	Convex	Concave

Designation of Welding and Allied Processes by Letters

GMAW-P	gas metal arc welding—pulsed arc	
GMAW-S	gas metal arc welding—short circuiting arc	
GTAC	gas tungsten arc cutting	
GTAW	gas tungsten arc welding	
GTAW-P	gas tungsten arc welding—pulsed arc	
HFRW	high frequency resistance welding	
HPW	hot pressure welding	
IB	induction brazing	
INS	iron soldering	
IRB	infrared brazing	
IRS	infrared soldering	
IS	induction soldering	
IW	induction welding	
LBC	laser beam cutting	
LBW	laser beam welding	
LOC	oxygen lance cutting	
MAC	metal arc cutting	
OAW	oxyacetylene welding	
OC	oxygen cutting	
OFC-A	oxyfuel gas cutting	
OFC-H	oxyhydrogen cutting	
OFC-N	oxynatural gas cutting	
OFC-P	oxypropane cutting	
OFW	oxyfuel gas welding	
OHW	oxyhydrogen welding	
PAC	plasma arc cutting	
PAW	plasma arc welding	
PEW	percussion welding	
PGW	pressure gas welding	
POC	metal powder cutting	
PSP	plasma spraying	
RB	resistance brazing	
RPW	projection welding	
RS	resistance soldering	
RSEW	resistance seam welding	
RSW	resistance spot welding	
ROW	roll welding	
RW	resistance welding	
S	soldering	
SAW	submerged arc welding	
SAW-S	series submerged arc welding	
SMAC	shielded metal arc cutting	
SMAW	shielded metal arc welding	
SSW	solid state welding	
SW	stud arc welding	
TB	torch brazing	

Typical Welding Symbols

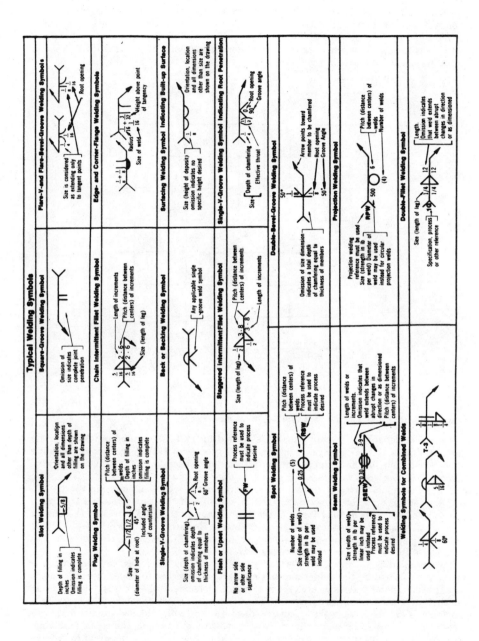

Slot Welding Symbol

Plug Welding Symbol

Single-V-Groove Welding Symbol

Flash or Upset Welding Symbol

Spot Welding Symbol

Seam Welding Symbol

Welding Symbols for Combined Welds

Square-Groove Welding Symbol

Chain Intermittent Fillet Welding Symbol

Back or Backing Welding Symbol

Staggered Intermittent Fillet Welding Symbol

Double-Bevel-Groove Welding Symbol

Projection Welding Symbol

Double-Fillet Welding Symbol

Flare-V- and Flare-Bevel-Groove Welding Symbols

Edge- and Corner-Flange Welding Symbols

Surfacing Welding Symbol Indicating Built-up Surface

Single-V-Groove Welding Symbol Indicating Root Penetration

ified CV machine with the fairly flat characteristic curve. The conventional CC machine can be used for manual welding and, under some conditions, for automatic welding. The CV machine is used *only* for continuous electrode wire arc-welding processes operated automatically or semiautomatically. These types of machines are best understood by comparing their respective volt-ampere characteristic output curves. This type of curve is obtained by loading the welding machine with variable resistance and plotting the voltage at the electrode and work terminals for each amperage output. Figure 3-46 shows an example of this curve.

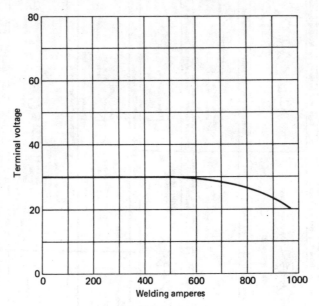

Figure 3-46   Voltage-current curve.

Conventional or Constant-Current Welding Machines

The conventional or CC welding machine is used for manual, covered (stick) electrode arc welding or SMAW, the gas tungsten (TIG) process or GTAW, carbon-arc welding (CAW), arc gouging, and stud welding (SW). It can be used for automatic welding with larger-sized electrode wire, but only with a *voltage-sensing* wire feeder.

The CC welder produces a volt-ampere output curve such as is shown in Fig. 3-47.

A brief study of the curve reveals that a machine of this type produces maximum output voltage with no load (zero current), and that as the load increases the output voltage decreases. Under normal welding conditions the output voltage is between 20 and 40 V. The open-circuit voltage is between 60 and 80 V; CC machines are available that produce either ac or dc welding power or both ac and dc.

When welding with covered electrodes on CC welding machines, the arc voltage is largely controlled by the welder and has a direct relationship to the arc length. As the arc length is increased (a long arc), the arc voltage increases. If the arc length is decreased (a short arc), the arc voltage decreases. The output curve shows that when the arc voltage increases (long arc), the welding current decreases, or when the arc voltage decreases (short arc), the welding current increases. Thus, without changing the machine setting, the welder can vary the current in the arc or welding heat to a limited extent by lengthening or shortening the arc.

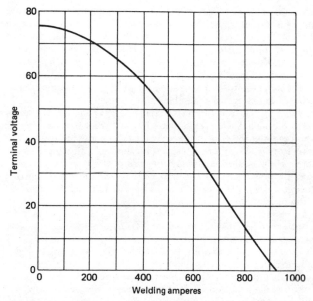

Figure 3-47 Voltage-current curve for constant-current machines.

CC machines can produce ac or dc welding power and can be rotating (generators) or static (transformers or transformer-rectifier) machines. The generator can be powered by a motor for shop use or by an internal-combustion engine (gasoline, LP gas, or diesel) for field use. Engine-driven welders can have either water- or air-cooled engines, and many of them provide auxiliary power for emergency lighting, power tools, etc.

Generator Welding Machines

On dual-control machines, normally generators, the slope of the output curve can be varied. The open-circuit, or no-load voltage is controlled by the fine-adjustment control knob. This control is also the fine welding-current adjustment during welding. The range switch provides coarse adjustment of the welding current. In this way, a soft or harsh arc can be obtained. With the flatter curve, and its low open-circuit voltage, a change in arc voltage will produce a greater change in output current. This produces a digging arc, preferred for pipe welding. With the steeper curve and its high open-circuit voltage, the same change in arc voltage will produce less of a change in output current. This is a soft, or quiet arc, useful for sheet-metal welding. In other words, the dual control, conventional, or CC welding generator allows the most flexibility for the welder. These machines can be driven by an electric motor or internal-combustion engine.

The CC, or drooping volt-ampere characteristic, machine can also be used for automatic welding processes. However, to use this type of welding machine, the automatic wire-feeding device must compensate for changes in arc length. This requires rather complex control circuits which involve feedback from the arc voltage or *voltage sensing*. This type of system is not used for the small-diameter electrode wire welding process.

Transformer Welding Machines

The transformer-type welding machine is the least expensive, lightest, and smallest of any of the different types. It produces alternating current for welding. The transformer takes power directly from the line, transforms it to the power required for welding and, by means of various magnetic circuits, inductors, etc., provides the volt-ampere characteristics proper for welding. The welding current output of a transformer may be adjusted in many different ways. The simplest method of adjusting output current is to use a tapped secondary coil on the transformer. This is a popular method employed by many of the limited-input, small welding transformers. The leads to the electrode holder and the work are connected to plugs, which may be inserted in sockets on the front of the machine in various locations to provide the required welding current. On some machines a tap switch is employed instead of the plug-in arrangement. In any case, exact current adjustment is not entirely possible.

On industrial types of transformer welding machines a continuous-output current control is usually employed. This can be a mechanical or electric control. The mechanical method involves moving the core of the transformer or moving the position of the coils within the transformer. The more advanced method of adjusting current output is by means of electric circuits. In this method the core of the transformer or reactor is saturated by an auxiliary electric circuit which controls the amount of current delivered to the output terminals. By adjusting a small knob, it is possible to provide continuous current adjustment from minimum to maximum output.

Although the transformer type of welder has many desirable characteristics, it also has some limitations. The power required for a transformer welder must be supplied by a single-phase system. This may create an unbalance in the power-supply lines, which is objectionable to most power companies. In addition, transformers have a rather low power-factor demand unless they are equipped with power-factor–correcting capacitors. The addition of capacitors corrects the power factor under load and produces a reasonable power factor which is not objectionable to electric power companies.

Transformer welding machines have the lowest initial cost, are the least expensive to operate, and require least space. In addition, ac welding power supplied by transformers reduces arc blow which can be troublesome on many welding applications.

Transformer-Rectifier Welding Machines

Some types of electrodes can be operated successfully only with dc power. A method of supplying dc power to the arc, other than using a rotating generator, is by adding a rectifier, an electric device which changes alternating current into direct current. Rectifier welding machines can be made to use a three-phase input. The three-phase input machine overcomes the line unbalance mentioned before.

In this type of machine the transformers feed into a rectifier bridge which then produces direct current for the arc. In other cases, where both alternating and direct current may be required, a single-phase ac transformer is connected to the rectifier. By means of a switch the welder can select either alternating or direct, straight- or reverse-polarity current for the welding requirement. In some types of ac-dc machines a high-frequency oscillator, plus water- and gas-control valves, are installed. This then makes the machine suited for gas tungsten arc welding as well as for manual coated-electrode welding.

The transformer-rectifier welding machines are available in different sizes and for single-phase or three-phase power supply. They may also be arranged for different primary voltages from the power line. The transformer-rectifier unit is more efficient electrically than the generator and provides quiet operation.

Multiple-Operator Welding System

This system uses a heavy-duty, high-current, and relatively high-voltage power source which feeds a number of individual-operator welding stations. At each welding station, a variable resistance is adjusted to drop the current to the proper welding range. Depending on the duty cycle of the welders, one welding machine can supply welding power simultaneously to a number of welders. The current supplied at the individual station has a drooping characteristic similar to the single-operator welding machines described above. The power source, however, has a CV output like those described next. The welding machine size and the number and size of the individual welding-current control stations must be carefully matched for an efficient multiple-operator system.

Constant-Voltage Welding Machines

A CV, sometimes called CP, or modified CV power source, is a welding machine that provides a nominally constant voltage to the arc regardless of the current in the arc. The characteristic curve of this type of machine is typified by the volt-ampere curve. *This type of machine can only be used for semiautomatic or automatic arc welding* using a continuously fed electrode wire. Furthermore, these machines are made to produce only direct current.

In continuous wire welding, the burn-off rate of a specific size and type of electrode wire is proportional to the welding current. As the welding current increases, the amount of wire burned off increases proportionally. This is graphically shown on the burn-off rate vs. current chart (Fig. 3-48). Thus, it can be seen that if wire were fed into an arc at

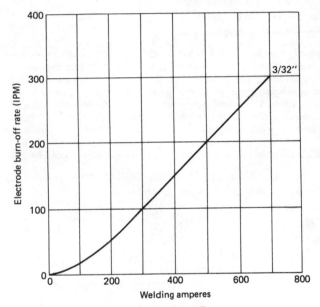

Figure 3-48   Chart showing burnoff rate vs. current.

a specific rate it would automatically draw a proportionate amount of current from a constant-voltage power source. The CV machine provides the amount of current required from it by the load imposed on it. The wire is fed into the arc by means of a constant-speed feed motor. This feed motor can be adjusted to increase or decrease the rate of wire feed. The system is inherently self-regulating. Thus, if the electrode wire were fed in faster, the current would increase. If it were fed in more slowly, the current would decrease automatically. The current output of the welding machine is thus set by the speed of the wire-feed motor. The voltage of the machine is regulated by an output control on the power source. Thus, only two controls maintain the proper welding current and voltage when the CV system is used.

The characteristic curves of CV machines have a slight inherent droop. This droop can be increased, or the slope made steeper, by various methods. Many machines have different tapes, or controls, for varying the slope of the characteristic curve. It is important to select the slope most appropriate to the process and the type of work being welded. CV machines can be either the generator type or the transformer-rectifier type.

Combination CV-CC Welding Machines

The most flexible type of welding machine is a combination type that can provide dc welding power with either a drooping or flat output characteristic volt-ampere curve by using different terminals and/or changing a switch. This type of welding machine is the most universal machine available. It allows the welder to use any of the arc-welding processes. The combination machine can be either a generator or a transformer-rectifier power source.

Specifying a Welding Machine

Selection of the welding machine is based on:

1. The process or processes to be used
2. The amount of current required for the work
3. The power available to the jobsite
4. Convenience and economic factors

These criteria plus information about each of the arc-welding processes indicate the type of machine required. The size of the machine is based on the welding current and duty cycle required. Welding current, duty cycle, and voltage are determined by analyzing the welding job and considering weld joints, weld sizes, etc., and by consulting welding procedure tables. The incoming power available dictates this fact. Finally, the job situation, personal preference, and economic considerations narrow the field to the final selection. The local welding equipment supplier should be consulted to help make the selection.

To order a welding machine properly, the following data should be given:

1. Manufacturer's type designation or catalog number
2. Manufacturer's identification or model number
3. Rated load voltage
4. Rated load amperes (current)
5. Duty cycle
6. Voltage of power supply (incoming)
7. Frequency of power supply (incoming)
8. Number of phases of power supply (incoming)

Welding-Machine Duty Cycle

Duty cycle is defined as the ratio of arc time to total time. For a welding machine, a 10-min time period is used. Thus for a 60 percent duty-cycle machine the welding load would be applied continuously for 6 min and would be off for 4 min. Most industrial-type CC (drooping) machines are rated at 60 percent duty cycle. Most CV (flat) machines used for automatic welding are rated at 100 percent duty cycle.

Figure 3-49, showing percent of working time vs. welding current, represents the ratio of the square of the rated current to the square of the load current multiplied by the rated duty cycle. Rather than work out the formula, use this chart. Draw a line parallel to the sloping lines through the intersection of the machine's rated current output and rated duty cycle. For example, a question might arise whether a 400-A, 60 percent duty cycle machine could be used for a fully automatic requirement of 300 A for a 10-min welding job. Line A shows this to be possible. It shows that the machine can be used at slightly over 300 A at a 100 percent duty cycle. Conversely, there may be a need to draw more than the rated current from a welding machine, but for a short period. Line B, for example, shows that the 200-A, 60 percent rated machine can be used at 250 A, providing the duty cycle does not exceed 40 percent (or 4 min out of each 10 min).

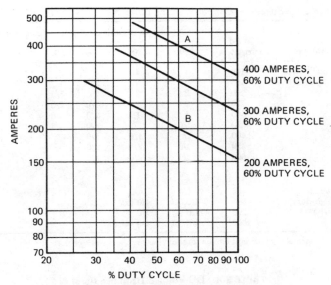

Figure 3-49 Percent of working time vs. welding current.

Use Fig. 3-49 to compare various machines. Relate all machines to the same duty cycle for a true comparison.

WELDING CABLE SELECTION

The size and length of the welding leads, welding cable and work cable, have a definite influence on the cost of welding. As the length of the leads is increased, their diameter should also be increased in order to avoid excessive voltage drop between the machine and the electrode and, particularly, to avoid wasted power as a result of the cables heating excessively.

To determine the power lost in the welding work leads, measure the voltage at the welding machine terminals. Then measure the voltage at the arc (meter connected between electrode holder and the work). Measure the welding current. The voltage loss in the leads equals the difference between the voltage at the terminals and that at the holder. Multiply this by the welding current, and the result is the power lost:

$$V \text{ (at terminal)} - V \text{ (at holder)} \times \text{welding current} = \text{power lost}$$

For example:

$$(35V - 32V) \times 250A = 750W \text{ lost}$$

Recommended Cable Sizes for Leads of Various Lengths

Table 3-3 shows cable sizes recommended for various lengths of leads. The footage shown includes the complete welding circuit—both welding lead and work lead combined. For example, the 60-ft column refers to two 30-ft leads.

DC Voltage Drop per 100 ft of Leads

Table 3-4 shows the voltage drop that will occur in a given length of cable of a given AWG size during welding with a given current value.

TABLE 3-3   Suggested Copper Welding Cable Size (AWG) Guide

Weld type	Weld current, A	Length of cable circuit, ft*—cable size					
		60	100	150	200	300	400
Manual (low-duty cycle)	100	4	4	4	2	1	1/0
	150	2	2	2	1	2/0	3/0
	200	2	2	1	1/0	3/0	4/0
	250	2	2	1/0	2/0		
	300	1	1	2/0	3/0		
	350	1/0	1/0	3/0	4/0		
	400	1/0	1/0	3/0			
	450	2/0	2/0	4/0			
	500	2/0	2/0	4/0			
Automatic (high-duty cycle)	400	4/0	4/0				
	800	4/0 (2)	4/0 (2)				
	1200	4/0 (3)	4/0 (3)				
	1600	4/0 (4)	4/0 (4)				

*Length of cable circuit equals total of electrode and work cables.

TABLE 3-4   DC Voltage Drop per 100 ft of Leads*

Welding current, A	Cable size (AWG)					
	2	1	1/0	2/0	3/0	4/0
50	1.0	0.7	0.5	0.4	0.3	0.3
75	1.3	1.0	0.8	0.7	0.5	0.4
100	1.8	1.4	1.2	0.9	0.7	0.6
125	2.3	1.7	1.4	1.1	1.0	0.7
150	2.8	2.1	1.7	1.4	1.1	0.9
175	3.3	2.6	2.0	1.7	1.3	1.0
200	3.7	3.0	2.4	2.0	1.5	1.2
250	4.7	3.6	3.0	2.4	1.8	1.5
300		4.4	3.4	2.8	2.2	1.7
350			4.0	3.2	2.5	2.0
400			4.6	3.7	2.9	2.3
450				4.2	3.2	2.6
500				4.7	3.6	2.8
550					3.9	3.1
600					4.3	3.4
650						3.7
700						4.0

*Figures in this table are for three-conductor cable. For four-conductor cable, reduce the ampere rating of each wire size by 20 percent.

When a cable is overheated, its life is shorter than when it is used without overheating. These figures assume that all connections are tight and that electrode holder and ground connections are in good condition.

For a higher amperage than given in the table, divide the load *equally* across two input cables of sufficient size to carry half the load.

To determine proper power cable wire size, consult the welding machine nameplate or data sheet for the amperage drawn at input line voltage. The data are based on NEC "Minimum Standards for Welding Equipment," Sec. 630.

WELDING DIFFERENT METALS

In order to produce a quality weld it is necessary to know the composition or analysis of each piece of metal that is to be welded. Once these properties or specifications are known, the proper filler metal can be selected so that the deposited weld metal will meet or exceed the mechanical properties of the base metal and have approximately the same composition and physical properties. The following two conditions must be fulfilled.

Base-Metal Properties

The base metal, or parts to be welded, must be known in order to determine their mechanical properties. The deposited weld metal must be selected to overmatch the mechanical properties of the base material. This can be done by selecting filler metals in accordance with some of the rules that follow.

Base-Metal Composition

The composition or analysis of the base metal, or metal to be welded, must be known. This can be determined if the specifications or trade name of the material is known. The filler metal to be used will then be selected in accordance with some of the rules that follow.

The exact selection procedure for filler metals varies somewhat depending upon the welding process that has been selected and, broadly, on the classification of metal that is to be welded.

Shielded-Metal Arc Welding

The shielded-metal–arc-welding process is the most popular and is commonly used for welding carbon steels, low-alloy steels, and stainless steels as well as for surfacing and other specialized applications. It is normally not used for welding aluminum, magnesium, titanium, and other "hard-to-weld" metals. Shielded-metal arc welding and the gas-shielded processes are used for welding the nickel alloys, copper alloys, high-strength steels, tool steels, and similar materials.

The following guidelines are to be used for selecting covered electrodes for welding carbon and low-alloy steels. These are related to the American Welding Society Filler Metal Specifications A5.1, "Carbon Steel Covered Arc Welding Electrodes," and A5.5, "Low Alloy Steel Covered Arc Welding Electrodes." The classification for these types of electrodes is shown in Table 3-5. The prefix letter "E" designates an electrode. The first two or three digits indicate tensile strength and other mechanical properties. The third (or fourth) digit indicates the welding position that can be used, and the last digit indicates usability of the electrode. The suffix letter, when used following the four- or five-digit classification, designates the composition of the deposit weld metal. This is normally used for low-alloy, high-strength electrodes and does not apply to the 60XX classification. The suffix letters and the nominal composition are shown in Table 3-6.

For exact data the filler metal specification should be consulted.

The operational factors relating to covered electrodes are as follows:

Welding Position

Electrodes are designed to be used in specific positions. The third (or fourth) digit of the electrode classification indicates the welding position that can be used. Match the electrode to the welding position that will be encountered.

Welding Current

Some electrodes are designed to operate best with direct current, others on alternating current. Some will operate on either alternating or direct current. The last digit indicates the welding current usability. Select the electrode to match the type of power source that will be used.

TABLE 3-5   AWS Classification System for Covered Mild and Low-Alloy Steel Electrodes

A. Prefix: E designates an electrode
B. First two or three digits: mechanical properties

Classification	Minimum tensile strength, lb/in^2 (MPa)	Minimum yield strength, lb/in^2 (MPa)	Minimum elongation, %
E60XX	62 000 (427)	50 000 (345)	22
E70XX	70 000 (483)	57 000 (393)	22
E80XX	80 000 (552)	67 000 (462)	19
E90XX	90 000 (621)	77 000 (531)	17
E100XX	100 000 (690)	87 000 (600)	16
E110XX*	110 000 (758)	97 000 (669)	15
E120XX*	120 000 (827)	107 000 (738)	14

C. Third (or fourth) digit: applicable welding positions
   EXX1X: flat, horizontal, vertical, and overhead
   EXX2X: flat and horizontal fillet
D. Last digit: electrode usability

Classification	Current†	Arc	Penetration	Covering slag	Iron powder, %‡
EXX10	dcep	Digging	Deep	Cellulose-sodium	0–10
EXX11	ac, dcep	Digging	Deep	Cellulose-potassium	0
EXX12	ac, dcen	Medium	Medium	Rutile-sodium	0–10
EXX13	ac, dcen, dcep	Soft	Light	Rutile-potassium	0–10
EXX14	ac, dcen, dcep	Soft	Light	Rutile-iron powder	25–40
EXX15	dcep	Medium	Medium	Low-hydrogen–sodium	0
EXX16	ac, dcep	Medium	Medium	Low-hydrogen–potassium	0
EXX18	ac, dcep	Medium	Medium	Low-hydrogen–iron powder	25–40
EXX20 and EXX22 (single pass)	ac, dcen, dcep	Medium	Medium	Iron oxide-sodium	0
EXX24	ac, dcen, dcep	Soft	Light	Rutile–iron powder	50
EXX27	ac, dcen, dcep	Medium	Medium	Iron oxide–iron powder	50
EXX28	ac, dcep	Medium	Medium	Low-hydrogen–iron powder	50
EXX48 (vertical down)	ac, dcep	Medium	Medium	Low-hydrogen–iron powder	25–50

*Low-hydrogen-type coating only.
†dcep = electrode positive—reverse polarity; dcen = electrode negative—standard polarity.
‡Iron powder percentage based on weight of the covering.

Thickness and Shapes of Base Metal

Weldments may include thick and heavy material of complicated design. The electrode selected should have maximum ductility to avoid weld cracking. Select the low-hydrogen types, EXX15, 16, 18, or 28.

Weld Design and Fit-up

Welding electrodes are designed with a digging, a medium, a soft, or a light-penetrating arc. The last digit of the classification indicates this usability factor. Deep-penetrating electrodes with a digging arc should be used when edges are not beveled or fit-up is tight. At the other extreme, light-penetrating electrodes with a soft arc are required when welding on thin material or when root openings are too wide.

TABLE 3-6 Chemical Composition of Deposited Weld Metal

Suffix	C	Mn	Si	Ni	Cr	Mo	V
			Weld metal composition, %*				
A1	0.12	0.60 or 1.00†	0.40 or 0.80†	——	——	0.40–0.65	——
B1	0.12	0.90	0.60 or 0.80†	——	0.40–0.65	0.40–0.65	——
B2L	0.05	0.90	1.00	——	1.00–1.50	0.40–0.65	——
B2	0.12	0.90	0.60 or 0.80†	——	1.00–1.50	0.40–0.65	——
B3L	0.05	0.90	1.00	——	2.00–2.50	0.90–1.20	——
B3	0.12	0.90	0.60 or 0.80†	——	2.00–2.50	0.90–1.20	——
B4L	0.05	0.90	1.00	——	1.75–2.25	0.40–0.65	——
B5	0.07						
‡	0.15	0.40–0.70	0.30–0.60	——	0.40–0.60	1.00–1.25	0.05
C1	0.12	1.20	0.60 or 0.80†	2.00–2.75	——	——	——
C2	0.12	1.20	0.60 or 0.80†	3.00–3.75	——	——	——
C3	0.12	0.40–1.25	0.80	0.80–1.10	0.15	0.35	0.05
D1	0.12	1.25–1.75	0.60 or 0.80†	——	——	0.25–0.45	——
D2	0.15	1.65–2.00	0.60 or 0.80†	——	——	0.25–0.45	——
‡							0.10 min.
G	——	1.00 min.	0.80 min.	0.50 min.	0.30 min.	0.20 min.	
M	0.10	0.60–2.25†	0.60 or 0.80†	1.40–2.50	0.15–1.50†	0.25–0.55†	0.05

*Compositions are maximum unless otherwise indicated.
†Amount depends on electrode classification.
‡A suffix is not applied to E60XX classification.

Service Conditions and Specifications

Weldments subjected to severe service conditions such as low temperature, high temperature, and shock loading, need special consideration. Select the electrode to match the base-metal properties including composition, ductility, and toughness. This is indicated by the toughness requirement of the specification. Usually low-hydrogen-type electrodes are required.

Production Efficiencies and Job Conditions

Certain electrodes are designed for high deposition rates but may be used only under certain position requirements. Where they can be used, select the high-iron powder types, the EXX24, 27, or 28 types. Other conditions may require some experimentation to determine the best electrode for the job, allowing for the most efficient production.

Carbon and low-alloy steel electrodes may be classed into four general groups:

F-1	High-deposition group	E6020, E7024, E6027, E6028
F-2	Mild-penetrating group	E6012, E6013, E7014
F-3	Deep-penetrating group	E6010, E6011
F-4	Low-hydrogen group	E6015, E7016, E7018, E6028

Electrodes in the same grouping operate and are run in the same general manner.

Electrode Selection for Constructional Steels

Welding procedure information is provided for most trade-name steels. This information is presented with a section for the constructional steels of each of the following steel companies: Armco Steel Corporation and its Sheffield Division, Bethlehem Steel Corporation, Great Lakes Steel Corporation, Inland Steel Corporation, Jones & Laughlin Steel Corp., United States Steel Corporation, Republic Steel Corporation, and Youngstown Sheet and Tube Company. The information given conforms to recommendations of that particular steel company. The trade name of each steel is given. In cases where the proprietary steel is also known to conform to certain ASTM specifications, this information is also given. These listings are not intended to list all the products of the companies named, but an attempt has been made to include the steels commonly used in welded fabrication at the time of this writing.

General Recommendations for Preheating and Electrode Selection

The recommendations are:

1. No welding should be done when the ambient temperature is below 0°F (−18°C). When the base metal's temperature is below 32°F, preheat the base metal to at least 70°F (0°C) (21°C) and maintain this minimum temperature during welding. Light sections require only local preheating, but heavy sections require general preheating. For structures, the American Welding Society specifies that the preheat be maintained on all surfaces of the plate within 3 in of the point of welding.

2. Electrodes that are not of the low-hydrogen type can be used to weld thinner sections of mild carbon steel when proper preheat temperatures are maintained. Low-hydrogen electrodes are recommended for thicker sections of steel and for low-alloy steel in all thickness ranges. When low-hydrogen electrodes are used, they must be thoroughly dry. They may be kept dry by storing them in a heated box and removing them from it immediately prior to use. Specific recommendations are found in the discussion of the Republic Steel Company's products. See Tables 3-12 and 3-13 in that discussion.

3. Any preheating indicated should be done prior to any tack welding as well as prior to the principal welding, and the temperature should be maintained as a minimum interpass temperature as welding proceeds.

4. When low-alloy steels are welded to lower-strength grades, select electrodes to match the strength of the lower-strength steel, but use welding practice suitable for the higher-strength steel.

Popular ASTM Constructional Steels

Popular steels are listed by the ASTM as follows:

A 36, Structural Steel

A 131, Structural Steel for Ships

A 201, Carbon-Silicon Steel Plates of Intermediate Tensile Ranges for Fusion Welded Boilers and other Pressure Vessels

A 212, High Tensile Strength Carbon-Silicon Plates for Boilers and Other Pressure Vessels

A 242, (Weldable grade) High Strength Low Alloy Structural Steel

A 283, Low and Intermediate Tensile Strength Carbon Steel Plates of Structural Quality

A 441, High Strength Low Alloy Manganese Vanadium Steel

Armco Steel Company (Including Sheffield Steels)

SSS-100, SSS-100A, and SSS-100B. Use low-hydrogen rods only. For high restraint, use 500°F (260°C) maximum preheat (400°F or 205°C maximum for SSS-100A and SSS-100B). It is important to keep heat input low to obtain fast cooling. Do not weave the electrode more than 2½ times the electrode diameter. Multipass welding employing the stringer bead technique should be used. Allow beads to cool below 250°F (120°C) (200°F or 93°C for SSS-100A and SSS-100B) before making additional passes. Use E110XX electrodes (E120XX in cases of high restraint). If SSS-100 is welded to a lower-strength material, low-hydrogen electrodes should be selected to match the strength of the weaker material. When postweld stress-relief heat treatment is applied to SSS-100 series steel weldments, the weld metal should not contain added vanadium. Stress-relief temperature is 1100°F (600°C). Fillet welds may be made with E90XX or E100XX. For joints where the weld metal is expected to provide yield and tensile strengths equal to that of the base metal, electrodes of the E110XX series are ordinarily employed.

Abrasion-Resistant SSS-100 Series Plates (321, 360 and 400 Bhn). The same procedures as outlined for the SSS-100 constructional alloys should be followed using techniques ordinarily employed on hardenable alloy steels. Weld metal with the lowest permissible strength often is selected to assure adequate ductility and toughness in the weld deposits. Where the hardness or toughness of a welded zone appears unsuited for service conditions, a postweld tempering may be done. However, the temperature should be limited to 800°F (430°C) to avoid lowering the overall hardness of the heat-treated plate.

Sheffield Hi-Strength A (ASTM A 242). Also called Armco High Strength #1. Use E60XX electrode and E70XX for multipass welding. Preheating or postheating is not necessary.

Sheffield Hi-Strength B (ASTM A 441). Also called Armco High Strength #5. Use E60XX or E70XX electrodes.

Sheffield Hi-Strength C (Grades 45, 50, 55, 60). Grade designations indicate minimum yield point. Grade 50 is called Armco High Strength #6; grade 45 is called Armco High Strength #7. No preheating is required for C steels. Use E70XX electrodes for grades 45 and 50. Use E80XX for grades 55 and 60. Use low-hydrogen electrodes for grades 55 and 60.

Sheffield Shef-Ten (ASTM A 440). This steel is primarily applicable to riveted and bolted structures. Steel under this specification can be satisfactorily welded under controlled conditions and proper procedures. No preheating is necessary in thicknesses up to and including ½ in. Preheat temperatures ranging from "hand warm" for thickness only slightly greater than ½ in to 350°F (180°C) for the greater thickness may be necessary. Preheating is required for cold metal.

Shef-Lo-Temp and Shef-Super-Low-Temp. Use low-hydrogen electrodes. Use E80XX-C1 or E80XX-C2. Weld deposits with high nickel content should be used for low-temperature applications.

Armco (Sheffield) Abrasion-Resisting Steel. Preheating to 300 to 400°F (150 to 205°C) and postheating or slow cooling after welding are recommended for this steel. Preheat also for tack welding. Use low-hydrogen electrodes of E100XX grade. If full-strength joints are not needed, lower-strength electrodes such as E70XX and E80XX may be used.

Aluminized Steel Type 1. Resistance welding is the best method for this steel. Corrosion resistance can be restored to weld areas by metallizing. Low-hydrogen E-6016 rods and 18-8 stainless steel rods can also be used. All slag and oxides must be removed to prevent corrosion. Weld areas should then be metallized.

Armco Hi-Strength #4R and 4S. No preheat required. Use E7018 or E7018-A1 electrode.

Armco QTC. Use low-hydrogen electrodes. E90XX and E100XX electrodes are suggested. Caution should be exercised to ensure dryness of both plate surfaces to be welded and flux coatings on electrodes.

Inland Steel Company

Hi-Steel (A 242). Use E-7018 or E-7018-A1 electrodes. Preheat of 200 to 500°F (93 to 260°C) may be required in the heavier sections or highly restrained weldments.

Tri-Steel (A 441). Use E-7018 or E-7018-A1 electrodes. Preheat may be required in the heavier sections or highly restrained weldments.

Hi-Man (A 440). Welding same as Tri-Steel. This steel is not recommended for welding. It can be welded when special care is taken, but is intended primarily for riveted or bolted construction.

INX Steels (45, 50 and 55). Use E-7018 or E-7018-A1 electrodes. Preheat may be required in heavier sections or highly restrained weldments.

INX Steels (60, 65 and 70). Use E-8018 or E-9018 electrodes. Preheat may be required when there is restraint.

Bethlehem Steel Company

V Steels. Low-hydrogen electrodes are recommended for most applications, particularly for thicknesses over ¾ in and for all thicknesses in V-55, V-60, and V-65. If the temperature of the shop or steel falls below 50°F (10°C) for conditions where no preheat is shown, preheat the steel to 100°F (38°C) before welding. Where V steels are to be welded to lower-strength grades, select electrodes to match the strength of the lower-strength steel, but employ welding practice for the higher-strength steel. See Table 3-7.

TABLE 3-7A Preheat Schedule for V Steels

Thickness, in	Type of electrode	Preheat schedule				
		V-45	V-50	V-55	V-60	V-65
<¾	Conventional	None	None	None	100°F	150°F
	Low-hydrogen	None	None	None	None	None
>¾ to ¾	Conventional	None	100°F	150°F	200°F	250°F
	Low-hydrogen	None	None	None	100°F	150°F
>¾ to 1½	Conventional	150°F	150°F	200°F	250°F	——
	Low-hydrogen	None	None	100°F	150°F	——
>1½ to 2	Conventional	200°F	250°F	300°F	——	——
	Low-hydrogen	150°F	150°F	200°F	——	——
>2 to 3	Conventional	300°F	300°F	350°F	——	——
	Low-hydrogen	200°F	250°F	300°F	——	——

TABLE 3-7B Recommended Electrodes for Manual Arc Welding of V Steels

Steel	Electrode
V-45	E60XX
	E70XX
V-50	E60XX
	E70XX
V-55	E70XX
V-60	E70XX
	E80XX
V-65	E70XX
	E80XX

Bethlehem Abrasion-Resisting Steel, Grade 235. A preheat and interpass temperature of 300 to 400°F (150 to 205°C) is recommended. Slow cooling is necessary after welding to prevent cracking. If the weld will be subject to impact, normalize at 1650°F (900°C). Use E10018-D2 or E10018-M rods. If preheat is impractical, use 309 or 312 stainless steel rods.

Mayari R(A242). Use E7018 or E7018-A1 rods. For greater corrosion resistance, use E8018-B2. For thicknesses from 1½ to 2 in (3.8 to 5 cm), preheat at 100°F (38°C); for thicknesses from 2 to 4 in (5 to 10 cm), preheat at 200°F (93°C).

Medium Manganese. Use E7018 or E7018-A1 rods.

Preheat temperature, °F (°C)

½″ to 1″	1 to 1½″	1½″ to 2″	2 to 4	over 4″
100 (38)	200 (93)	200 (93)	300 (150)	350+ (177+)

Manganese Vanadium.   Use E7018 or E7018-A1 rods.

Preheat temperature, °F (°C)

To 1″	1 to 1½″	1½ to 2″	2 to 4	over 4″
50	100	200	300	350+
(10)	(38)	(93	(150)	(177+)

Great Lakes Steel Corporation

GLX-45-W, GLX-55-W, GLX-50-W, GLX-60-W.   These steels are classified according to their yield strength. Thus, GLX-45-W indicates a mild carbon steel with a yield strength of 45,000 lb/in^2 minimum. Electrodes recommended are those that would be used at the same carbon grade, except that the class should overmatch the parent metal in strength. For multipass welds, electrodes of the AWS E70XX class are satisfactory.

Steel	Electode
N-A-XTRA 80	Use E9015 or E9018
N-A-XTRA 90	Use E10015 or E10018
N-A-XTRA 100	Use E11015 or E11018
N-A-XTRA 110	Use E12015 or E12018

Equivalent electrodes for alternating current or iron-bearing coatings are satisfactory.

X-A-R 15 and X-A-R 30.   The high hardenability of these steels requires care in welding. Preheat and interpass temperature may be required for highly restrained weldments and when welding material more than ¾ in (19 mm) thick. E10018-M, E10018-D2, or E12018-M electrodes should be used.

Jones & Laughlin Steel Corp.

J & L Cor-Ten.   Use AWS E7018 or AWS E7018-A1 electrodes. Where greater corrosion resistance is required use AWS E8018-B2. No preheat is required.

Jalloy-S-90.   Preheat is not normally required unless a unique stress-distribution condition arises. Use AWS E10XX low-hydrogen series.

Jalloy-S-100.   Preheat same as for Jalloy-S-90. Use AWS E110XX low-hydrogen series.

Jalloy-S-110.   Preheat same as for Jalloy-S-90. Use AWS E120XX low-hydrogen series.

Jalloy-AR-280, Jalloy-AR-320, Jalloy-AR-360, and Jalloy-AR-400.   Must use a low-hydrogen rod. Preheat is not normally required unless a unique stress-distribution condition arises. Use AWS E100XX, E110XX, or E120XX. If preheat is required, 200 to 400°F (93 to 205°C) should be used.

JLX-45-W, JLX-50-W, JLX-55-W, JLX-60-W.   Use E7018 or E7018-A1 electrodes. No preheat is required.

J & L Nickel-Copper-Titanium High-Strength Forming Steel.   Preheat is required only if a code must be met. Use an E60XX electrode (E6010 or E6012). For highly stressed conditions, an E70XX rod may be required.

Jalten No. 1, No. 2, and No. 3. Use AWS E60XX electrode. An E70XX rod may be required for highly stressed conditions. Use 400 to 600°F (205 to 316°C) preheat in critical cases.

Speed Case and Speed Treat. Must use low-hydrogen rods, normally E70XX, but may have to go higher depending on design. A short arc is desirable with proper current an important factor. Preheat may be required.

Speed Alloy. Must use low-hydrogen electrode. Preheat to 500°F (260°C) before welding. Normalizing or full annealing is recommended after welding. A short arc with proper current is essential. Use AWS E100XX rods of low-hydrogen series.

United States Steel Corporation

USS Spec	ASTM Spec
Man-Ten	A440
Tri-Ten	A441
Cor-Ten	A242

The general rules on preheat and electrode selection apply unless superseded by a specific note for that steel.

Cor-Ten Steel (A 242). When the welded area in multiple-pass welds is to simulate or approach the color of Cor-Ten steel after atmospheric exposure, electrodes containing 2¼ percent nickel (AWS E8016-C3) or 3½ percent nickel (AWS E8016-C2) are suggested for shield-metal arc welding. For single-pass welding, carbon steel electrodes of the E60XX and E70XX group are satisfactory. Low-hydrogen electrodes are preferred for thickness greater than ½ in (12 mm). A preheat temperature above the minimum given may be required for highly restrained welds.

For shielded-metal arc welding the practices in Table 3-8 are suggested. The use of an austenitic stainless steel electrode of the E309, E310, or E312 class is recommended for joining USS Cor-Ten to austenitic stainless steel.

Tri-Ten Steel (ASTM A 441 and A 242). Low-hydrogen electrodes are preferred for manual arc welding. USS Tri-Ten can also be welded to austenitic stainless steel by using electrodes of the E309, E310, or E312 class. For suggested welding practices see Table 3-9.

Man-Ten and Man-Ten A440 Steels. Man-Ten A440 is intended for riveted or bolted structures. Man-Ten is considered weldable under carefully controlled conditions (see Table 3-10). Low-hydrogen electrodes are preferred for USS Man-Ten Steel. Man-Ten A440 is intended for riveted or bolted structures. However, when attachments are made by welding, the minimum preheat temperatures in Table 3-10 are necessary.

TABLE 3-8 SMAW Welding Practices for Cor-Ten Steel

Electrode	Thickness, in	Suggested min. preheat or interpass temp, °F (°C)
Low-hydrogen type (E60 or E70, 16, 18 or 28) of ASTM A-233	<1 incl*	50 (10)
	>1 to 2 incl	100 (38)
	>2 to 5 incl	200 (93)
Other than low-hydrogen type (E60 or E70 group) of ASTM A-233	to ½	50 (10)
	>½ to 2 incl	200 (93)
	>2 to 5 incl	300 (150)

*incl = inclusive.

TABLE 3-9   Welding Practices for Tri-Ten Steel

Electrode	Thickness, in	Min preheat or interpass temp, °F (°C)
Low-hydrogen type (E60 or E70, 15, 16, 18 or 28) of ASTM A-233	to 1 incl*	50 (10)
	> 1 to 2 incl	100 (38)
	>2	200 (93)
Other than low-hydrogen type (E60 or E70 group) of ASTM A-233	to ½ incl	50 (10)
	>½ to 1 incl	100 (38)
	>1 to 1½ incl	200 (93)
	>1½ to 2 incl	250 (120)
	>2	300 (150)

*incl = inclusive.

TABLE 3-10   Suggested Welding Practices for Man-Ten and Man-Ten A440 Steels

Electrodes	Thickness, in	Man-Ten A440 min preheat or interpass temp, °F (°C)	Man-Ten min preheat or interpass temp, °F (°C)
Low-hydrogen type (E60 or E70 15, 16, 18 or 28) of ASTM A-233	to ⅜ incl	50 (10)	50 (10)
	>⅜ to ½ incl*	100 (38)	50 (10)
	>½ to 1 incl	200 (93)	100 (38)
	>1 to 2 incl	300 (150)	200 (93)
	>2	300 (150)	300 (150)
Other than low-hydrogen type (E60 or E70 group) of ASTM A-233	to ½ incl	NR*	200 (93)
	>½ to 1 incl	NR	300 (150)
	>1 to 2 incl	NR	300 (150)
	>2	NR	NR

*incl = inclusive; NR = not recommended.

A-R Steel. Low-hydrogen electrodes of the ASTM-AWS E9015, E9016, E9018, E10016, and E10015 grades should be used for welding USS A-R steel. If the weldment is to be postweld heat-treated, the E10016 or other electrodes containing more than 0.05 percent vanadium should not be used for welding. Recommended preheating for A-R steel is

Thickness, in	Minimum preheat or interpass temperature, †F (°C)
<½ inclusive	300 (150)
>½ to 2 inclusive	400 (205)

Postweld heat treatment for 1 h of thickness is recommended (1025°F or 555°C).

Ex-Ten Steel. Welding operations with Ex-Ten occasionally result in residual stresses or increased hardness of sufficient magnitude to require postweld heat treatment, particularly in the higher strength grades. See Table 3-11. USS Ex-Ten steel can be welded to structural carbon steel or to other USS high-strength steels using an ordinary mild or low-alloy steel electrode as desired, subject to the practices suggested for the grade. Ex-Ten steel can be welded to austenitic stainless steel by using electrodes of the E309, E310, or E312 class.

TABLE 3-11   Suggested Welding Practice for Ex-Ten Steel

		Low-hydrogen electrode		Non-low-hydrogen electrode	
Grade	Thickness, in	Min preheat or interpass temp, °F (°C)	Electrode	Min preheat or interpass temp, °F (°C)	Electrode
70	to ⅜ incl*	50 (10)	E8015, 16, 18	NR*	NR
60	to ⅜ incl	50 (10)	E70, 15, 16, 18 or 28	50 (10)	E70XX
	>⅜ to ¾ incl	50 (10)	E70, 15, 16, 18 or 28	200 (93)	E70XX
	>¾ to 1¼ incl	100 (38)	E70, 15, 16	NR	NR
50	to ⅜ incl	50 (10)	E60 or E70, 15, 16, 18 or 28	50 (10)	E60XX or E70XX
	⅜ to ¾ incl	50 (10)	E60 or E70, 15, 16, 18 or 28	100 (38)	E60XX or E70XX
	¾ to 1 incl	50 (10)	E60 or E70	200 (93)	E60XX or E70XX
	1 to 1½ incl	100 (38)	E60 or E70, 15, 16, 18 or 28	200 (93)	E60XX or E70XX
	1½ to 2 incl	100 (38)	E60 or E70, 15, 16, 18 or 28	NR	NR
	>2	200 (93)	E60 or E70, 15, 16, 18 or 28	NR	NR
42	to ¾ incl	50 (10)	E60 or E70, 15, 16, 18 or 28	50 (10)	E60XX or E70XX
	¾ to 1 incl	50 (10)	E60 or E70, 15, 16, 18 or 28	100 (38)	E60XX or E70XX
	1 to 1½ incl	50 (10)	E60 or E70, 15, 16, 18 or 28	200 (93)	E60XX or E70XX
	1½ to 2 incl	50 (10)	E60 or E70, 15, 16, 18 or 28	NR	NR
	>2	150 (66)	E60 or E70, 15, 16, 18 or 28	NR	NR

*incl, inclusive; NR, not recommended.

Republic Steel Company

Steel		Electrode
Republic	70	E80XX, E70XX, E10018, E10016
Republic	65	E70XX, E80XX, E10016, E10018
Republic	50	E60XX, E70XX, E7016-E7018
Republic	M-1	E60XX, E7016 or 18
Republic	M-2	Same as M-1
Republic	A-441	E60XX, E70XX
X-45-W		E60XX, E70XX
X-50-W		E60XX, E70XX
X-55-W		E60XX, E70XX
X-60-W		E60XX, E70XX

See Tables 3-12 and 3-13 for preheat temperatures.

Welding Stainless Steel

Selection of Covered Electrodes for Welding Stainless Steels

In order to properly select the electrode for welding stainless steels or, more correctly, corrosion-resisting steels, it is necessary to know and understand the numbering system used. This numbering system, established by the American Iron and Steel Institute

TABLE 3-12   Preheat for Non-Low-Hydrogen Rods, °F

Rod	To ½ in	½ to 1 in	1 to 1½ in	1½ to 2 in	2 to 4 in	Over 4 in
70	NR	NR	NR	NR	NR	NR
65	NR	NR	NR	NR	NR	NR
50	50	50	50			
M-1	50	50	200	200		
M-2	50	50	200	200		
A-441	NR	NR	NR	NR	NR	NR
X-45-W	50	150	150	200	NR	NR
X-50-W	50	150	150	200	NR	NR
X-55-W	150	250	250	250	NR	NR
X-60-W	200	250	250	300	NR	NR

*NR, not recommended.

TABLE 3-13   Preheat for Low-Hydrogen Rods, °F

Rod	To ½ in	½ to 1 in	1 to 1½ in	1½ to 2 in	2 to 4 in	Over 4 in
70	50	100	150			
65	50	50	100			
50	50	50	50	150	200	
M-1	50	50	50	50	150	
M-2	50	50	50	50	150	
A-441	50	200	300	300	300	350+
X-45-W	50	50	50	150	200	
X-50-W	50	50	50	150	250	
X-55-W	50	100	100	200	300	
X-60-W	50	100	150			

(AISI), is based on the composition of the stainless steel, i.e., type 308, type 312, etc. These are three-digit numbers which classify the steel according to its metallurgical structure. Stainless steels are sometimes known and identified according to their principal alloying element, such as 18/8, 25/20, and so on.

Iron is the main element of all stainless steels. However, to make it corrosion-resistant, chromium must be present in an amount of 11.5 percent or more. The addition of chromium to iron provides a fine film of chromium oxide which forms on the surface and acts as a barrier to further oxidation, rust, or corrosion. The addition of nickel in the proper ratio results in a stainless steel series referred to as chrome-nickel types. They all contain a percentage of nickel and are nonmagnetic. The addition of nickel increases the corrosion resistance, ductility, electric resistance, impact properties, and fatigue resistance.

There are three basic classes of stainless steel which are grouped according to their metallurgical microstructure. They are known as the austenitic, martensitic, and ferritic types. The properties of these three classes of stainless steel differ and require different welding electrodes and procedures.

Electrodes for welding stainless steels are identified by the AISI three-digit number following the prefix letter "E" and followed by the usability classification, normally 15 or 16. The usability classification can also be followed by a letter such as "L" indicating "low carbon." All stainless steel electrodes have a low-hydrogen-type coating. The EXXX-15 type indicates a lime coating which is normally used with direct current. The EXXX-16 type uses titanium-base coating and can be used with alternating or direct current. Type 16 is a smoother-running electrode. However, type 15 may be the best for out-of-position welding. The mechanical properties are not specified for stainless steel electrodes since the deposit weld metal will be nearly identical to the base metal and will have physical properties normal to the composition of the metal deposit. Table 3-14 gives the proper electrodes for welding the different types of stainless steel.

TABLE 3-14   Electrode Selection for Welding Stainless Steels

AISI No.	Carbon	Manganese	Silicon	Chromium	Nickel	Other elements	Hobart Electrode No.
			Chemical analyses of stainless steels, %				
			Austenitic				
201	0.15 max.	5.5–7.5	1.0	16.0–18.0	3.5–5.5	N_2 0.25 max.	308
202	0.15 max.	7.5–10.	1.0	17.0–19.0	4.0–6.0	N_2 0.25 max.	308
301	0.15 max.	2.0	1.0	16.0–18.0	6.0–8.0	——	308
302	0.15 max.	2.0	1.0	17.0–19.0	8.0–10.0	——	308
302B	0.15 max.	2.0	2.0–3.0	17.0–19.0	8.0–10.0	——	308
303	0.15 max.	2.0	1.0	17.0–19.0	8.0–10.0	S 0.15 min.	308DC
303Se	0.15 max.	2.0	1.0	17.0–19.0	8.0–10.0	Se 0.15 min.	308DC
304	0.08 max.	2.0	1.0	18.0–20.0	8.0–12.0	——	308
304L	0.03 max.	2.0	1.0	18.0–20.0	8.0–12.0	——	308L
305	0.12 max.	2.0	1.0	17.0–19.0	10.0–13.0	——	308
308	0.08 max.	2.0	1.0	19.0–21.0	10.0–12.0	——	308
309	0.20 max.	2.0	1.0	22.0–24.0	12.0–15.0	——	309
309S	0.08 max.	2.0	1.0	22.0–24.0	12.0–15.0	——	309
310	0.25 max.	2.0	1.50	24.0–26.0	19.0–22.0	——	310
310S	0.08 max.	2.0	1.50	24.0–26.0	19.0–22.0	——	310
314	0.25 max.	2.0	1.5–3.0	23.0–26.0	19.0–22.0	——	310DC
316	0.08 max.	2.0	1.0	16.0–18.0	10.0–14.0	Mo 2.0/3.0	316
316L	0.03 max.	2.0	1.0	16.0–18.0	10.0–14.0	Mo 2.0/3.0	316L
317	0.08 max.	2.0	1.0	18.0–20.0	11.0–15.0	Mo 3.0/4.0	317
321	0.08 max.	2.0	1.0	17.0–19.0	9.0–12.0	Ti 5 × C min.	347
347	0.08 max.	2.0	1.0	17.0–19.0	9.0–13.0	Cb + Ta 10 C min.	347
348	0.08 max.	2.0	1.0	17.0–19.0	9.0–13.0	Ta 0.10 max.	347
			Martensitic				
403	0.15 max.	1.0	0.5	11.5–13.0	——	——	410
410	0.15 max.	1.0	1.0	11.5–13.5	——	——	410
414	0.15 max.	1.0	1.0	11.5–13.5	1.25–2.5	——	410
416	0.15 max.	1.25	1.0	12.0–14.0	——	S 0.15 min.	410DC
416Se	0.15 max.	1.25	1.0	12.0–4.0	——	Se 0.15 min.	410DC
420	Over 0.15	1.0	1.0	12.0–14.0	——	——	410
431	0.20 max.	1.0	1.0	15.0–17.0	1.25–2.5	——	430
440A	0.60–0.75	1.0	1.0	16.0–18.0	——	Mo 0.75 max.	——
440B	0.75–0.95	1.0	1.0	16.0–18.0	——	Mo 0.75 max.	——
440C	0.95–1.2	1.0	1.0	16.0–18.0	——	Mo 0.75 max.	——
			Ferritic				
405	0.08 max.	1.0	1.0	11.5–14.5	——	Al 0.1/0.3	410
430	0.12 max.	1.0	1.0	14.0–18.0	——	——	430
430F	0.12 max.	1.25	1.0	14.0–18.0	——	S 0.15 min.	430DC
430FSe	0.12 max.	1.25	1.0	14.0–18.0	——	Se 0.15 min.	430DC
446	0.20 max.	1.50	1.0	23.0–27.0	——	N 0.25 max.	309

Welding Nonferrous Metals

The Gas Tungsten Arc Process and Gas Metal Arc Process

The gas-shielded arc-welding processes are more popular for welding nonferrous metals. When using these processes it is best to select a welding electrode having a composition similar to the metal being welded. Unfortunately, electrodes are not available in every conceivable composition and, therefore, charts showing recommended electrodes for different types of metals are available. A guide for the choice of filler metals for welding aluminum is given in Table 3-15. This chart can be used for either GMAW or GTAW. Similar charts are available for welding magnesium. However, for welding with nickel-base or copper-base electrodes, the manufacturer of the base metal or manufacturers of nonferrous electrodes should be consulted.

TABLE 3-15 Guide to the Choice of Filler Metal for Aluminum

	Base metal	Casting alloys 43, 355, 356	214, A214, B214, F214	6061, 6062, 6063, 6151	5456	5454	5154, 5254 (1)	5086, 5356	5083	5052, 5652 (1)	5005, 5050	3004, CLAD 3004	1100, 3003, CLAD 3003	1060
A	1060	ER4043	ER4043	ER4043	ER5356 (3), (5)	ER4043	ER4043	ER5356 (3), (5)	ER5356 (3), (5)	ER4043	ER1100 (3)	ER4043	ER1100 (3)	ER1260
B	1100, 3003 CLAD 3003	ER4043	ER4043 (5)	ER4043 (5)	ER5356 (3), (5)	ER4043 (5)	ER4043 (5)	ER5356 (3), (5)	ER5356 (3), (5)	ER4043 (5)	ER4043 (5)	ER4043 (5)	ER1100 (3)	
C	3004, CLAD 3004	ER4043 (5)	ER4043 (5)	ER4043 (2)	ER5356 (5)	ER5356 (3), (5)	ER5356 (5)	ER5356 (5)	ER5356 (5)	ER5356 (3), (5)	ER4043 (5)	ER4043 (4), (5)		
D	5005, 5050	ER4043 (5)	ER4043 (5)	ER4043 (2)	ER5356 (5)	ER4043 (5)	ER5356 (3), (5)	ER5356 (5)	ER5356 (5)	ER4043 (5)	ER4043 (4), (5)			
E	5052, 5652 (1)	ER4043 (5)	ER4043 (2)	ER4043 (2)	ER5356 (5)	ER5356 (2), (3)	ER5356 (2)	ER5356 (5)	ER5356 (5)	ER5652 (2), (3)				
F	5083	ER5356 (3), (5)	ER5356 (5)	ER5356 (5)	ER5183 (5), (6)	ER5356 (5)	ER5356 (5)	ER5356 (5)	ER5183 (5), (6)					
G	5086, 5356	ER5356 (3), (5)	ER5356 (5)	ER5356 (5)	ER5356 (5)	ER5356 (5)	ER5356 (5)	ER5356 (5)						
H	5154, 5254 (1)	ER5356 (3), (5)	ER5356 (2)	ER5356 (2), (3)	ER5356 (5)	ER5356 (2)	ER5254 (2)							
I	5454	ER5356 (3), (5)	ER5356 (2)	ER5356 (2), (3)	ER5356 (5)	ER5554 (5)								
J	5456	ER5356 (3), (5)	ER5356 (5)	ER5356 (5)	ER5556 (5), (6)									
K	6061, 6062, 6063, 6151	ER4043 (5)	ER5356 (2), (3)	ER5356 (2), (3)										

	214, A214, B214, F214		
L		ER4043 (5)	ER5356 (2)
M	43,355, 356	ER4043 (4)	

Source American Welding Society.

(1) Base-metal alloys 5652 and 5254 are used for hydrogen peroxide service. ER5254 filler metal is used for welding both alloys for low-temperature service (150F and below). ER5652 filler metal is used for welding 5652 for high-temperature service (150F and above).

(2) ER5154, ER5254, ER5183, ER5356, ER5554, and ER5556 may be used. In some cases, they provide: (1) improved color match after anodizing treatment, (2) highest weld ductility, and (3) higher weld strength. ER5554 is suitable for elevated temperature service.

(3) ER4043 may be used for some applications.

(4) Filler metal with the same analysis as the base metal is sometimes used.

(5) ER5356, ER5183, or ER5556 may be used.

(6) ER5356 is the third choice.

chapter 1-4

Metal Resurfacing by Thermal Spraying

by
Richard J. DuMola
Materials Engineer
METCO Inc.
Westbury, New York

GLOSSARY

Listed below are the terms most commonly used in the thermal spray industry together with their definitions.

Abrasive   A hard material such as sand, aluminum oxide, steel grit, or silicon carbide used to clean and roughen a surface.

Base metal   The part or substrate to be resurfaced.

Blasting   The process in which an abrasive material is propelled, usually by air pressure, onto a surface to effect both cleaning and roughening.

Coating   The spray material which is applied to the base metal.

Deposition rate   The amount of material which adheres to the base metal per unit of time.

Bond or bond strength   A measure of how well a sprayed coating has adhered to the base metal.

Thermal spraying A group of processes which involve the melting and accelerating of an atomized spray of particles onto a surface forming a solid coating.

INTRODUCTION

The application of a coating by the thermal spray process is an established industrial method for resurfacing metal parts. The process is characterized by the simultaneous melting and transporting of the spray material, usually a metal or ceramic, onto the surface of the part to be coated. The spray material is propelled in the form of fine molten droplets which, upon striking the part, flatten, solidify, and adhere by a mechanical and metallurgical interaction. Each applied layer of spray material bonds tenaciously to the previously deposited layer. The process is continued until the desired coating thickness is achieved.

Thermal spraying can be used to apply a coating to machine element or structural parts to satisfy any one of the following broad requirements:

1. To repair worn areas on parts damaged in service
2. To restore dimension to mismachined parts
3. To increase the service life of a part by optimizing the surface physical properties

In addition to satisfying any one of these broad requirements, thermal spraying can be a cost-effective repair procedure when compared to the high cost of replacing worn or mismachined parts and the economic losses incurred as a result of machine downtime.

The primary advantages of thermal spraying over other methods of metal resurfacing are the wide range of chemically different materials which can be sprayed, the high coating deposition rate which allows thick coatings to be applied economically, and the portability of the spray equipment.

PHYSICAL PROPERTIES OF COATINGS

Thermally sprayed coatings are composed of individual particles of the spray material alloyed and mechanically interlocked together to form a solid coating. In general, there is only limited metallurgical bonding between the coating and the base metal. The coating adheres primarily by a mechanical anchoring mechanism. To ensure adequate bonding of the coating, the base metal must be free from oil or dirt contamination and should be roughened by machining or blasting.

Sprayed coatings are harder and more wear-resistant than cast or wrought alloys of the same material. The increased properties are due to fine oxides and a combination of work-hardening and rapid quenching of the spray particles upon impact with the base metal. Rapid quenching causes hard metastable phases to form.

Some degree of porosity is present in all sprayed coatings and results from the presence of air gaps between the spray particles. Typically, thermally sprayed coatings are 80 to 95 percent as dense as cast or wrought alloys of the same material. In applications where the coating is used as a bearing surface, the porosity helps to retain lubricating oil and gives the coating a degree of self-lubricity. In corrosion applications where it is necessary to protect the base metal, the coating should be sealed with an epoxy or aluminum vinyl paint to close off the pores.

The surface texture or roughness of thermally sprayed coatings is coarser than cast or wrought surfaces, and a subsequent finishing operation such as sanding, machining, or grinding is often required before the resurfaced part can be placed into service. Thermally sprayed coatings are generally not as machinable as wrought or cast alloys of the same chemistry due to the presence of oxides in the coating. Because tool wear is greater, sprayed coatings should be machined with the most abrasion-resistant carbide cutting tools available.

MATERIAL SELECTION

The first step in the selection of a thermally sprayed coating for a specific in-plant application is to define the coating function. For example, if a badly worn shaft is to be repaired by thermal spraying, then the desired coating function is increased wear resistance. Table 4-1 lists the most common coating functions encountered, one or two typical application areas, and the appropriate thermal spray materials which satisfy each coating function. It should be noted that for each coating function a number of spray materials are indicated. In order to pinpoint the best material for the application, secondary considerations such as equipment available, coating thickness required, material available, and final method of finishing should be evaluated.

PROCESS DESCRIPTION

Thermal spraying consists of four basic processes: wire flame spraying, powder flame spraying, arc wire spraying, and plasma arc spraying. Of these processes wire flame spraying and powder flame spraying are the most widely used in industry. (See Fig. 4-1.)

Both wire and powder flame spraying utilize the heat generated by the combustion of an oxygen fuel flame (typically oxyacetylene, oxypropane, or oxyhydrogen) to melt the spray material. A wire flame spray gun pulls the wire into the combustion flame by means of either a self-contained, variable-speed, air-driven turbine or an electric motor. A high-pressure stream of air both constricts the combustion flame and atomizes the molten tip of the wire, forming a spray of metal. A powder flame spray gun operates by feeding a fine powder into the combustion flame by a combination of suction and gravity. The powder is both melted and propelled by the combustion flame. The equipment typically required for either wire or powder flame spraying is listed in Table 4-2.

Both arc wire spraying and plasma arc spraying utilize an electric arc rather than a

TABLE 4-1 Coating Function and Material Selection*

Coating function	Applications	Material selection
Adhesive wear	Bearings Piston guides	Aluminum bronze Phosphor bronze Tin-base babbitts
Abrasive wear	Shafts Couplings Cutting blades	Hardfacing alloys Molybdenum Carbon steel Stainless steel Tungsten carbide
Atmospheric and saltwater corrosion	Exposed steel structures	Aluminum Zinc
High-temperature oxidation	Exhaust mufflers Annealing pans	Nickel chromium Aluminum
Restoration of dimensions	Mismachined parts and castings	Carbon steels Stainless steel Nickel alloys Aluminum

*Adapted from *Handbook of Coating Recommendations,* METCO Inc., New York, 1970.

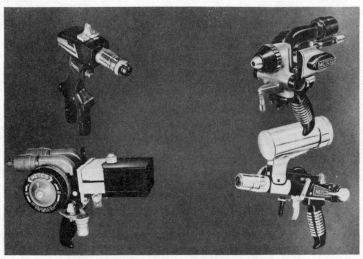

Figure 4-1   Thermal spray equipment. (*METCO Inc.*)

combustion flame to melt the spray material. Arc wire spraying is becoming increasingly important because it yields lower operating costs and higher deposition rates than wire flame spraying. Plasma arc spraying is utilized primarily where optimum coating density and overall quality are critical requirements as in aircraft applications.

Thermal spraying will generate airborne dust and metal fumes to varying degrees, depending on the material being sprayed. As a result, always provide adequate ventilation in the spray area to ensure operator safety. In cases where sufficient ventilation is not possible, spray operators should be equipped with dust masks or auxiliary ventilation equipment such as an exhaust hood or exhaust booth.

Each thermal spray process produces an intense bright light generated by either an electric arc or combustion flame. Eye protection as provided by dark glasses or an approved welding helmet is required.

In cases where the spray material is known to be toxic or suspected of containing potentially toxic elements, the recommendations of the thermal spray equipment manufacturer should be followed.

TABLE 4-2   Equipment Requirements

Wire or powder flame spray gun
Oxygen and fuel gas cylinders
Two-stage oxygen regulator
Two-stage acetylene regulator
Gas flowmeters (oxygen, fuel gas, and air)
Compressed air (30 ft³/min at 75 lb/in²)*
Gas hoses
Wire straightener (optional)

*85 m³/min at 500,000 N/m²

BIBLIOGRAPHY

Ballard, W. E.: *Metal Spraying and the Spray Deposition of Ceramics and Plastics,* 4th ed., Charles Griffin, London, 1963.

Burns, R. M., and W. W. Bradley: *Protective Coatings for Metals,* 3d ed., Reinhold, New York, 1967.
Ehrhardt, R. A., and A. Mendizza: "Sprayed Metal Coatings" in *Metals Handbook,* 1948 ed., American Society for Metals, Metals Park, Ohio, 1948.
Ingham, H. S., and A. P. Shepart: *METCO Flame Spray Handbook,* vols. I and II, METCO Inc., New York, 1964.
Longo, F. N.: *Handbook of Coating Recommendations,* METCO Inc., New York, 1972.
"Thermal Spray Terms and Their Definitions," Booklet AWS C2.9-70, American Welding Society, New York, 1970.

chapter 1-5

Structural Adhesives

by
Weldon M. Scardino, P.E.
Consulting Engineer
Centerville, Ohio

GLOSSARY*

Adherend A body which is held to another body by an adhesive.

Adhesive, contact An adhesive that is apparently dry to the touch and which will adhere to itself instantaneously upon contact.

Adhesive, heat-setting An adhesive that requires a temperature above 87°F (31°C) to set it.

*Reprinted with permission from the *Annual Book of ASTM Standards*, Part 22. Copyright American Society for Testing and Materials, 1916 Race Street, Philadelphia, PA 19103.

Adhesive, pressure-sensitive A viscoelastic material which in solvent-free form remains permanently tacky. Such material will adhere instantaneously to most solid surfaces with the application of very slight pressure.

Adhesive, room-temperature setting An adhesive that sets in the temperature range of 68° to 86°F (20° to 30°C).

Bond strength The unit load, applied in tension, compression, flexure, peel, impact, cleavage, or shear, required to break an adhesive assembly with failure occurring in or near the plane of the bond.

Catalyst A substance that markedly speeds up the cure of an adhesive when added in minor quantity compared with the amounts of the primary reactants.

Creep The dimensional change with time of a material under load, following the initial instantaneous elastic or rapid deformation.

Cure To change the physical properties of an adhesive by chemical reaction.

Failure, adhesive Rupture of an adhesive bond such that the separation appears to be at the adhesive-adherend interface.

Failure, cohesive Rupture of an adhesive bond such that the separation appears to be within the adhesive.

Fillet Adhesive which fills the corner or angle where two adherends are joined.

Flow Movement of adhesive during the bonding process, before the adhesive is set.

Hardener A reacting substance or mixture that promotes or controls the curing reaction.

Joint, lap Joint made by overlapping and bonding two adherends.

Polymerization A chemical reaction whereby monomer molecules link together to form larger ones.

Post cure To expose an adhesive to additional cure (usually heat), following initial cure.

Primer A coating applied to a surface, prior to the adhesive, to improve the bond performance.

Storage life Time during which adhesive can be stored and still be suitable for use (also called *shelf life*).

Structural adhesive Adhesive used for transferring required loads between adherends exposed to service environments typical for the structure involved.

Surface preparation Physical and/or chemical preparation of an adherend to make it suitable for adhesive bonding.

Thermoplastic A material that will repeatedly soften when heated and harden when cooled.

Thermoset A material that will undergo or has undergone a chemical reaction by the action of heat, catalyst, etc., leading to a relatively infusible state.

Time, assembly The time interval between the spreading of the adhesive on the adherend and the application of pressure or heat, or both, to the assembly.

Working life The period of time during which an adhesive, after mixing with catalyst, solvent, or other compounding ingredients, remains suitable for use.

INTRODUCTION

The use of adhesives, both structural and nonstructural, is increasing rapidly. The advent of the epoxies made much of this possible, although they are not the only types of structural adhesives. Adhesives are used for simple repairs to equipment having low operational stresses and also in many applications requiring strength equal to or greater than that of the parent material or part. For instance, epoxy repairs to concrete and many types of reinforced plastics as well as metals can be expected to reinforce these base materials, so that later breakage will be outside the adhesive joint. Adhesives can now be used to replace rivets, screws, and welding.

The modern plant engineer is often faced with a variety of fastening problems to be solved with a minimum of cost and time. The many adhesives available today can often meet all of these requirements best; i.e., they solve the problem, and are quick and inexpensive to use. Frequently, adhesives furnish the *only* solution to problems where mechanical fasteners simply cannot be used.

In addition to frequently furnishing the best or only solution to a fastening problem, adhesives usually provide structural joints with the best fatigue resistance. This is true for two reasons: the ability of the adhesive to damp, absorb, and distribute stresses, and the fact that no holes are needed for fastening. Most failures of fastened parts originate at fastener holes in the form of fatigue or stress cracks which get progressively larger. A bonded joint eliminates this problem.

However, adhesives are not a panacea for all fastening problems. The successful use of adhesives requires a careful choice of adhesive properties to suit the adherends, proper joint design, proper surface preparation, proper cure temperature and pressure, and an adhesive system that can resist the environmental exposure to which the bonded joint will be subjected.

CLASSIFICATIONS

Hot-Melt Adhesives

These are thermoplastic polyvinyl acetate and other chemicals either melted at point of application or melted in a container or gun and dispensed as a liquid at the point of application. The hot melts have long been used in shoe construction and packaging and are now being used in carpet splicing and many other applications.

Advantages

Rapid application; fast setting; low cost; indefinite shelf life; moisture and solvent resistance; ability to bond to many surfaces; nontoxic; one part, no mixing.

Disadvantages

Poor heat resistance [softening point of typical materials is 104 to 212°F (40 to 100°C)]; special equipment needed for application; low shear strength [500 to 1000 lb/in² (3450 to 6900 kPa) although some can go as high as 3500 lb/in² (24 129 kPa)]; poor creep resistance.

Cyanoacrylate Adhesives

The cyanoacrylate adhesives, or *miracle glues,* are well known for their advertising aimed at the home market. They cure by the action of the minute amount of moisture present on most surfaces.

Advantages

Very fast setting and curing (seconds or minutes); ease of application; special equipment not needed; good tensile and shear strength; one part, no mixing.

Disadvantages

Work best on nonpermeable surfaces; closely mated surfaces needed for most types; poor impact strength; poor solvent, moisture, and heat resistance; limited to small bonding areas because of fast set; careful handling needed (can bond skin almost instantly).

Anaerobic Adhesives

These one-part materials cure when confined in a joint that excludes oxygen. Long used for thread sealing and locking, they are available in many strength variations and viscosities. Some are true structural adhesives. Primer is required on some surfaces to counteract contamination or to "activate" the surface.

Advantages

One component, no mixing; ease of application; can be somewhat gap-filling as compared to cyanoacrylates; tough, good impact resistance; temperature operating limit as high as 300 to 450°F (149 to 232°C); fast cure; long shelf life, easy to store.

Disadvantages

Sometimes require primer; not for all substrates; for nonpermeable surfaces only.

Acrylic Adhesives

Acrylic adhesives are thermosetting liquids and pastes. Early-generation adhesives have been improved and have advantages in a variety of operations. They are fast-setting, easy to mix and use, and they can tolerate a broad range of bond-joint gaps. Their strength is close to that of the epoxies, and some have very good moisture resistance when cured. For special applications, the curing agent can be applied to one side in advance as a primer.

Advantages

Fast cure at room temperature; high-strength peel and shear; good moisture resistance obtainable; good gap filling; tolerant of surface contamination; with primer, can be used on permeable surfaces.

Disadvantages

Strong, objectionable odor; limited container open time.

Epoxies

Probably the most widely used industrial structural adhesives, these 100 percent solid thermosetting materials have been used for several decades and are constantly being improved. The peel strength and impact resistance have been improved by modification with elastomers. Recently, improved durability versions for *primary* structural bonding of aircraft parts have been developed and "qualified." Nylon-epoxy and epoxy-polyamide adhesives have good impact strength and elongation, but poor moisture and heat resistance. The phenolic epoxies have excellent heat resistance, but poor elongation and impact resistance.

Epoxy adhesives can be obtained as one-part liquids, pastes, or films and as two-part pastes. The one-part adhesives have a much shorter shelf life than the two-part adhesives and usually must be stored at 0 to 40°F (−18 to 4°C). The two-part adhesives generally cure at room temperature and require only contact pressure. However, they have limited heat and moisture resistance. The one-part epoxy adhesives fall in two general classes—those that cure at 250°F (121°C) and are usable at 180°F (82°C) and those that cure at 350°F (177°C) and are usable to 300 to 350°F (149 to 177°C). An adhesive that cures at an intermediate temperature [300°F (149°C)] is also available. The film versions are particularly suitable for large and intricate surface areas and are easy to apply. The pastes and liquids can be applied by spatula, roller, or spray. Some epoxies can also be obtained in powdered form for electrostatic and hot-dip application. For best results, prefitting and proper pressure to obtain bond lines of approximately 4 to 10 mils (0.1 to 0.25 mm) or 6 mils (0.15 mm) are necessary for most adhesives. Bond-line "shimming" with glass beads or filaments is sometimes used with paste-type adhesives to obtain the optimum bond thickness. The film adhesives are easier to control in this respect, as most are on a "carrier" which helps to perform this task. As a general rule, the higher the cure temperature, the greater the heat and humidity resistance and the lower the peel and impact strength of an epoxy. Only a few room-temperature adhesives can match the properties of the heat-cured epoxies.

Frequently, room-temperature-curing epoxies that set rapidly use an amine curing agent, which can be toxic. These fast-setting epoxies are also quite brittle. The room-temperature-curing adhesives that use polyamide curing agents have good impact resistance.

Epoxies can be modified by fillers to give various properties, including electric conductance. Some typical uses of epoxies involve everything from attaching jewels to jewelry to bonding aircraft primary structures. Almost all helicopter rotor blades are epoxy-bonded laminated metal or composite structures. Most aircraft leading edges, trailing edges, spoilers, rudders, and elevators are epoxy-bonded assemblies. These can be metal or composite (either graphite-epoxy or fiberglass-epoxy).

Advantages

Obtainable in many different forms: liquids, paste, films, or powder—one-part, or two-part; variability of cure time, cure temperature, and pot life; inert to most solvents and moisture (depending on type); wide variety of shear, peel, and stiffness properties available; strongest and most dependable for most structural bonding applications.

Disadvantages

Require careful surface preparation—usually chemical treatment; some equipment needed for mixing and dispensing paste types; require proper attention to cure process; limited shelf life for one-component types—require refrigerated storage; require primer for best environmental results.

Phenolics

Phenolic adhesives cover a wide range of types: water dispersions for bonding wood, one-component heat-curable liquids; solutions, films, or powders and liquid solutions plus catalyst for lower-temperature cures. Because of their heat-resistant properties, various combination adhesives are formulated, such as nitrile phenolics, vinyl phenolics, and epoxy phenolics. Nitrile-phenolic adhesives have been very successfully used to *bond-seal* aircraft integral fuel tanks.

Advantages

Heat resistance; wide range of forms available; humidity and solvent resistance.

Disadvantages

Brittle; emission of volatiles upon curing; not good gap fillers; lower strength than epoxies; high pressure required during cure, 100 to 200 lb/in^2 (690 to 1380 kPa).

Polyurethanes

Polyurethane adhesives are 100 percent solid materials in a liquid form. They may be one or two components and can be room-temperature or heat cured. Their use is limited.

Advantages

Nonbrittle; good for joining dissimilar materials; good for low-temperature (cryogenic) applications—strengths to 5000 lb/in^2 (34 500 kPa) at 100°F (-73°C); high peel susceptibility.

Disadvantages

Moisture-sensitive, before and after cure; poor lap shear at room and elevated temperature.

Silicones

The silicones are 100 percent solid pastes or liquids that usually cure at room temperature. Most one-part materials cure by reacting with moisture in the air. Because of their high cost and low shear strength, their use is confined to specialty applications where their unique properties are necessary.

Advantages

One- or two-part; wide range of temperature use [-80 to $500°F$ (-62 to $260°C$)]; high peel and impact strength; moisture resistance; good for dissimilar adherend joining; one part, easy to apply; solvent resistance.

Disadvantages

High cost; poor lap shear strength—usually not more than 500 to 600 lb/in² (3.45 to 4.14 MPa); poor creep resistance; for best adhesion usually requires primer; some can be reverted by heat and moisture.

Polyimides

Polyimide adhesives may be obtained as thermoplastic liquids and as one- or two-part thermosetting films or pastes. Their use is limited mostly to applications requiring their high-temperature resistance. Curing takes a long time at $500°F$ ($260°C$) and usually requires a high-temperature postcure.

Advantages

High-temperature resistance.

Disadvantages

High cost; limited production availability; high cure temperatures; emission of volatiles during cure by some types; brittle (low impact strength).

JOINT DESIGN

Most adhesives are strong in tension or shear strength but very weak in cleavage and/or peel strength (see Fig. 5-1). All cleavage and peel forces should be eliminated from adhesive joints to the greatest extent possible. Good joint design should also allow for the maximum possible bond area and mechanical locking as well as adhesive bonding. The simple lap joint of Fig. 5-2 can be improved in many ways: by increasing the thickness of each adherend or by making a type of double-lap shear joint. Both joints reduce the peel and cleavage forces at the edge of the bond joint caused by the eccentricity of the adherends. Other typical bonded joints are shown in Fig. 5-2. Figure 5-3 shows both good and bad ways to bond flexible adherends.

PREPARATION FOR BONDING

Surface Preparation

Surface preparation is the most critical step in the adhesive bonding process. Unless a satisfactory surface preparation is accomplished, the bond will fail adhesively and unpredictably at the adherend-primer interface. With proper surface preparation, bonds can be accomplished that will allow any failure to be cohesive in nature, thus realizing the predicted strength of the adhesive, and/or primer combination. The proper surface prep-

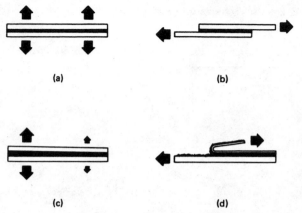

(a)

(b)

(c)

(d)

Figure 5-1 Stresses on bonded joints.

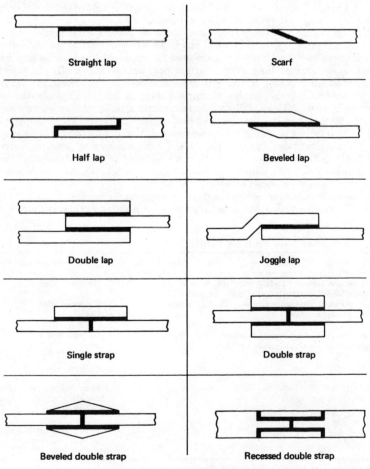

Straight lap

Scarf

Half lap

Beveled lap

Double lap

Joggle lap

Single strap

Double strap

Beveled double strap

Recessed double strap

Figure 5-2 Types of bonded joints.

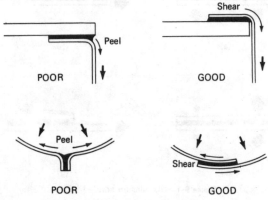

Figure 5-3   Joints for flexible adherends.

aration is a key factor, not only in the initial strength of a bonded joint but, even more importantly, *in its long-term environmental resistance.*

Surface-preparation methods must, as a minimum, remove oil, grease, or any coating whose bond strength to the adherend is apt to be less than that of the adhesive bond. Simple abrasion and/or solvent wiping are used for some metallic and plastic adherends. However, on most metals these simple surface preparations are usually not sufficient to obtain good adhesion or long-term environmental resistance. For metals, usually the preferred surface treatment is one that chemically removes organic contamination and the original oxide coating. This is followed by immediate priming or bonding or else, as for aluminum, the building up of a controlled oxide layer. ANSI/ASTM D 2093[5] and ANSI/ASTM D 2651-79[5] contain detailed recommended methods for preparation of various plastic and metal surfaces. (See Tables 5-1 and 5-2.) When using epoxy adhesives that cure at 250°F (121°C) or lower, the preferred treatment for most aluminum alloys is a patented phosphoric acid anodizing process described in ARP 1524[6] or the nonpatented, hand-applied version described in ARP 1575.[7]

The most widely used aluminum surface preparation for bonding is the FPL (Forest Products Laboratory) etch. This is described in ANSI/ASTM D 2651-79.[5] Recent

TABLE 5-1   Surface Treatments for Plastics*

Material	Procedure
Cellulose acetate, cellulose acetate butyrate, cellulose nitrate, cellulose propionate, ethyl cellulose, methyl styrene, polycarbonate, polystyrene, vinyl chloride, polymethylmethacrylate	1. Methanol wipe 2. Sand 3. Methanol wipe 4. Dry
Epoxy, polyester, phenolic, urea-formaldehyde, diallyl phthalate, melamine, nylon, polyurethane	1. Acetone wipe 2. Sand 3. Dry wipe 4. Acetone wipe
Polyethylene, polypropylene, chlorinated polyether, polyformaldehyde	Must be treated with sulfuric acid–dichromate solution (see ASTM D 2093)[5]
Teflon	Must be treated with sodium-naphthalene-complex, a strong oxidizer (see ASTM D 2093)[5]

*Abstracted from ANSI/ASTM D 2093 (Ref. 5)

TABLE 5-2 Surface Treatments for Metals*

Material	Procedure
Aluminum	1. Phosphoric acid anodizing per ARP 1524[6] or ARP 1575[7] or 2. FPL (Forest Products Lab.) etch:* Degrease, hot alkaline cleaner, hot sulfuric acid–sodium dichromate etch, rinse, and dry.
Carbon steel and stainless steel	Degrease, followed by mechanical abrasion–vapor blasting, sand blasting, etc.
Titanium, magnesium, copper	Require chemical treatment.

*Abstracted from ANSI/ASTM D 2093 (Ref. 5)

changes in the process have been incorporated in the document; these changes improve the environmental resistance of bonds prepared using this process.

Environmental Considerations

In selecting an adhesive for a particular application, one of the most important considerations is the environment or surroundings to which the adhesive joint will be subjected. Of course, the force acting on the joint is of prime consideration, and the adhesive joint must be capable of carrying the maximum expected load (without excessive creep) and amount of fatigue or cyclic stresses. Cyclic stresses, particularly slow ones, are much more damaging to an adhesive joint than a steady stress. The adhesive selected for a particular application must be able to resist these loads and stresses not only initially but also after exposure to the most severe environmental factors to be encountered during the life of the adhesive joint. Heat and humidity are usually the most damaging environmental factors for most bonded joints. Thermal-expansion stresses created between dissimilar materials having widely different coefficients of thermal expansion, e.g., a plastic-to-metal bond joint, require low-modulus (nonbrittle) adhesives for best performance. Other deleterious factors are solvents and ultraviolet or other energy. Always choose an adhesive that is resistant to these factors; do not plan on coating the adhesive joint with some "protective" coating which can possibly crack or eventually become permeable to solvents or moisture.

Primers

For many adhesives, a primer is not merely desirable but absolutely essential in order to obtain maximum bond strength and environmental durability. Usually, the adhesive manufacturer recommends a compatible adhesive primer. The primer performs several functions. The primer, being less viscous than the adhesive, wets the adherend surface and adheres to it better than does the adhesive. Corrosion-resistant primers used particularly with epoxies and aluminum adherends contain chromates that leach out and protect the adherend from corrosion. Corrosion-inhibited adhesive primers (CIAPs) are essential to environmentally durable aluminum bonding.

SPECIFICATIONS AND STANDARDS

The principal sources of specifications and standards are those issued by the government and by industry "voluntary" associations. The government documents are free. The industry documents may be purchased both individually or (for ASTM) in related groups or "books." Consult their indexes for the ASTM and AMS specifications, and the DoDISS (Department of Defense Index of Specifications and Standards) for all Government specifications and standards. The DoDISS also includes many industry specifications and standards that have been officially accepted for use by the Government.

Federal and Military Specifications Standards and Handbooks, Naval Publications and Forms Center, 5801 Tabor Avenue, Philadelphia, PA 19120, (215)697-2179.
SAE (AMS) Specifications, Society of Automotive Engineers, 400 Commonwealth Drive, Warrendale, PA 15096, (412)776-4841.
ASTM Specifications, American Society for Testing and Materials, 1916 Race Street, Philadelphia, PA 19103, (215)299-5400.

REFERENCES AND SOURCES OF INFORMATION

1. *Adhesives Desk-Top Data Bank,* 3d ed., The International Plastics Selector, Inc., San Diego, 1980–1981.
2. Thrall, E. W., and R. W. Shannon: *Adhesive Bonding of Aluminum Alloys for Aircraft,* Marcel Dekker, New York, 1982.
3. *Adhesives Red Book, Directory of the Adhesives Industry.* Communication Channels, Inc., Atlanta, 1982.
4. *Adhesives in Modern Manufacturing,* Society of Manufacturing Engineers (SAE), Dearborn, 1970.
5. *ASTM Annual Book of Standards,* part 22, American Society for Testing and Materials, 1982.
6. ARP 1524, "Aerospace Recommended Practice, Surface Preparation and Priming of Aluminum Alloy Parts for High Durability Structural Bonding, Phosphoric Acid Anodizing," 1978.
7. ARP 1575, "Aerospace Recommended Practice, Surface Preparation and Priming of Aluminum Alloy Parts for High Durability Structural Adhesive Bonding, Hand Applied Phosphoric Acid Anodizing," 1979.

chapter 1-6

Diagnostic Instrumentation for Machinery Maintenance

by
J. Mark Gilstrap
Technical Manager, Mechanical Engineering Services
Bently Nevada Corp.
Minden, Nevada

INTRODUCTION

In recent years, the level of sophistication of rotating machinery has indeed increased. Modern machines are operating at higher speeds, temperatures, pressures, and flow rates. It is common practice to continuously monitor the *process variables* (pressure, temperature, flow) in a machine in order to assess total plant operation. The techniques used to monitor these variables are described elsewhere in this Handbook.* When

*See Chap. 11-1 of Rosaler and Rice: *Standard Handbook of Plant Engineering*, 1983.

machinery reliability is considered, it is also necessary to monitor the onstream mechanical condition of that machinery. Of the available parameters to accurately determine mechanical integrity, the measurement of machine *vibration and position* characteristics has consistently proved to be a powerful tool.[1] Furthermore, when a monitor system indicates a problem with a machine, the early diagnosis of the malfunction is not only desirable, but often mandatory.

Successful vibration monitoring and analysis require an intimate familiarity with the types of measurements, transducer characteristics and applications, plus the capabilities and limitations of the diagnostic instrumentation. Ultimately, the vibration signals must be reduced to a hard-copy data format for engineering evaluation and presentation to plant maintenance, operations, and/or management personnel. This chapter will deal with the types of monitors and the various instruments used for machinery diagnosis and will describe the data presentations available.

MONITORING SYSTEMS

Monitoring systems provide the function known as *preventive maintenance*. The objective is to prevent, or at least limit, machinery damage due to mechanical or process malfunctions. It is economically favorable to monitor critical machinery.[2] The benefits are increased plant and personnel safety, decreased maintenance costs, reduced spare parts, lower insurance rates, and those factors associated with the highest dollar values, reduced downtime and increased plant availability. Thus, those machines which operate in a dangerous process, involve high capital expenditures, have high maintenance costs, or are critical to a major part of a plant's operation should have the most complete monitoring systems available. Semicritical machines would justify a less-sophisticated monitoring system, and noncritical machines may justify only a scanning-type microprocessor system or periodic measurements with portable instrumentation.

Standard monitor systems usually provide two alarm levels, an *alert* or first level of warning and a *danger* or shutdown alarm. Monitors are also equipped with a circuit and indicating light which ensures the proper operation of the transducer and field wiring. The American Petroleum Institute has published the most complete specification on monitoring systems for rotating machinery.[3] Recently, computer systems have been used increasingly in plants with many major machine trains. Computers can provide the absolute-limit alarms found in conventional monitors, but can also monitor rate of change and significant bandwidths (increases or decreases) and can compare one measured variable with others on the same machine. Computers are excellent for routine data storage, trending, alarm sequencing, and data comparison in a machine-upset condition.

Careful attention must be given to the selection of the appropriate measurement transducers for a monitor system.[4] With respect to machine vibration, transducers are naturally divided into two groups: (1) velocity transducers and accelerometers which measure machine housing motion, and (2) proximity probes which measure shaft motion. The objective of monitoring is to protect the machine against the most likely potential malfunctions. The most common malfunctions should then be evaluated in terms of shaft-related or machine-housing-related mechanisms. Then the transducer selected should be the type that will be the most reliable indicator of a change in the mechanical condition of the machine. Of course, on some machines, shaft vibrations may be transmitted to some extent to the housing and vice versa. However, the extent of transmissibility is a function of the *mechanical impedance* of a machine and is highly variable. Transducer selection should not be compromised for the sake of ease of installation; a monitor system can only be as reliable as the initial measurement.

It must be recognized that monitoring is not analyzing. A monitor may indicate when a machine is in trouble and may even indicate the severity of the problem. But the identification of the malfunction requires the use of additional instrumentation for machinery analysis.[5] Also, monitors are great averagers of events. For routine data acquisition on machines in *normal* operation, monitor amplitude levels should be recorded periodically or even continuously. But monitor readings can be enhanced greatly by the use of specialized instruments which measure more than amplitude alone. In some cases where transducer selection for monitoring is somewhat compromised, it will be beneficial to have permanently installed unmonitored transducers on the same machine. These transducers can then be measured periodically with portable monitors or specialized instrumentation.

INSTRUMENTATION FOR DIAGNOSIS

If a monitor system indicates a change in the running condition of a machine, then additional instruments are usually necessary to qualify and quantify the significance of that change. Some of these instruments are misnamed "analyzers." An instrument does not *analyze* data, it merely reduces it into a format for subsequent analysis by the cognizant technician or engineer. These so-called analyzers cannot compete with the human brain. Sound engineering judgment is still an absolute necessity.

Instruments used for machinery diagnostics aid the engineer in the function of *predictive maintenance*. Once a monitor or a periodic measurement of amplitude indicates a machine is in trouble, then these instruments must be used to predict: (1) what the mechanical or process malfunction is, (2) how long the machine can operate safely before a shutdown becomes necessary, (3) what operational changes may be made on-line to reduce the severity of the trouble and allow the machine to operate longer until an orderly shutdown can be planned, and (4) upon shutdown, what corrective action is necessary to restore the mechanical integrity of the machine.

It should be obvious that before any machinery analysis begins, the transducers providing the measurement signals should be verified for accuracy. Assuming that the basic measurement system has been installed and maintained properly, a complete recalibration should not be necessary. A few fundamental checks with a voltmeter and an oscilloscope will suffice. Besides, an unnecessary system recalibration will steal valuable time from the analysis program. Once proper transducer operation is verified, it is generally recommended that the more basic instruments be used in the initial analysis. Many of the most frequently occurring machine problems can be detected through the use of very basic instruments. On the other hand, the very sophisticated instruments may overlook some of the basic machinery data and yield misleading conclusions. For example, rotor unbalance, misalignment, internal rubs, and oil whirl can usually be recognized on an oscilloscope display.

Keyphasor

A Keyphasor is a function which provides a once-per-turn reference signal to shaft rotation.[6] The Keyphasor signal can be from a proximity probe observing a shaft discontinuity (such as a keyway) or from an optical transducer observing a light (or dark) spot on the shaft. When the Keyphasor signal pulse is generated, the shaft is at a known rotational position; the shaft event (discontinuity or bright spot) is directly under the Keyphasor probe. This signal can then be compared with respect to time to the shaft vibration signal. With this information, the rotational position of the shaft is related to the angular position of the shaft in its vibration cycle at any point in time. The Keyphasor function is not used alone with any instrument, except for a tachometer to measure shaft rotative speed. But, in conjunction with vibration signals, it is used with such instruments as oscilloscopes and tracking filters. Specific use of the Keyphasor will be explained in following sections.

Voltmeter

The voltmeter is a basic tool for transducer and monitor-system calibration and can also be a valuable machinery diagnostic aid when proximity probes are used. The dc output signal from a proximity transducer represents the *average* distance between the probe and the shaft. This information is processed directly by axial-thrust position monitors. However, for radially mounted probes, the standard vibration monitor usually only processes the ac or dynamic signal. The dc output of radial probes represents the average shaft centerline position within the bearing clearance. This information can be used to detect such parameters as: (1) shaft lift-off from the bottom of the bearing during machine start-up, (2) shaft attitude angle, (3) eccentricity ratio, (4) oil film thickness, (5) internal or external misalignment, (6) other steady-state unidirectional shaft preloads, (7) bearing wear, and (8) electrostatic discharge.[7] A plot of the average shaft centerline position relative to the bearing clearance during a machine start-up can be made from hand-logged probe gap voltages or from a magnetic tape recording with an XY plotter. A typical plot is shown in Fig. 6-1.

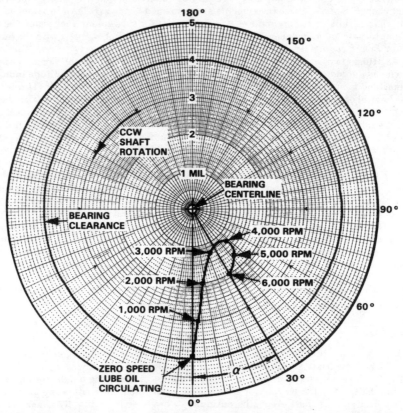

Figure 6-1   Change in average radial shaft centerline position during start-up. Diametral bearing clearance = 8 mils.

Oscilloscope

The oscilloscope is another basic tool for ensuring proper transducer and monitor-system calibration. It is also *the most fundamental* vibration-analysis instrument. Although other instruments are called *real-time analyzers,* the scope comes the closest to meeting the definition. The scope can display the raw, composite vibration signal from any type of transducer and can also show the dc signal of a proximity probe. It is most useful when used with a Keyphasor signal, and an oscilloscope camera can be used for hard-copy documentation of the displays.

The oscilloscope has two basic presentation forms: (1) time-base or waveform, and (2) orbit (Lissajous) or XY displays. The time-base display uses the sweep function of the scope to observe vibration amplitude with respect to time. Usually the sweep time is very short—three or four shaft revolutions—and the Keyphasor pulse can be superimposed on the time-base waveform to indicate the period of one shaft revolution. In the time-base mode, several vibration characteristics can be measured directly on or determined from the display: (1) vibration amplitude, (2) vibration phase angle, (3) vibration frequency, and (4) shaft rotative speed. When the Keyphasor signal is superimposed on the vibration signal(s), then it is easy to determine synchronous and nonsynchronous vibrations. It is particularly informative to use a scope with the output of a filter instrument and compare it with the unfiltered waveform.

The second form of presentation is the orbit or Lissajous. This is possible on scopes which have an XY or left-vs.-right channel function. The orbit is the dynamic motion of the shaft centerline as measured in two perpendicular planes. Usually an orbit is from two shaft-observing proximity probes at 90°, but it can be a casing orbit from XY seismic transducers. Essentially, all of the information (including the Keyphasor) available in the time-base display is also available, or can be confirmed, in the orbit display. In addition, shaft orbit *shape* has become a fundamental form of vibration analysis. Very unique orbit shapes indicate problems such as misalignment, oil whirl, and rubs, including the actual location of the rub. Unbalance simply shows an orbit which grows from a small circle or ellipse to a larger one. A Keyphasor with an orbit can be used very effectively for balancing.[8]

The oscilloscope can be twice as useful an instrument when it is used with a scope camera. The camera provides documentation for historical purposes. Most scope manufacturers have a camera and mounting bezel available with their scopes. It is also easy to use the camera in a multiple-exposure mode to obtain more information on one photograph than would normally be provided in a single oscilloscope display. A photo that looks like Fig. 6-2 can be obtained two ways: (1) a double exposure on a four-trace scope, or (2) a quadruple exposure on a dual-trace scope.

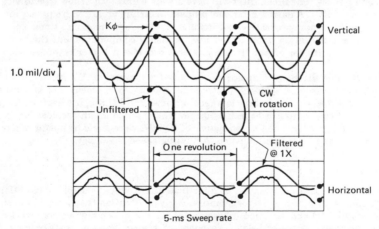

Figure 6-2 Oscilloscope tracing showing filtered and unfiltered waveforms and orbits.

Filters

Almost every instrument with the exceptions of the voltmeter, oscilloscope, and recording devices used for vibration analysis employs some sort of filter. This means that a certain vibration frequency or band of frequencies will be isolated and either (*a*) discarded or (*b*) retained for analysis. So, by definition, a filter will throw away some information. The general rule to follow is to be sure the filter is not discarding any information which may be important.

The most common type of filter is the bandpass filter. This filter will pass (or retain) a certain frequency (or very narrow band of frequencies) and discard all others. The bandpass filter is used to identify specific frequencies present in a composite vibration signal and to measure the amplitudes at each frequency. High-pass and low-pass filters are opposites. A high-pass filter discards all information below a certain cutoff frequency while a low-pass filter discards all information above a certain frequency. These filters are used to eliminate high- or low-frequency noise or unrelated vibrations at the ends of the frequency spectrum. A band-reject or notch filter is the opposite of a bandpass filter. It is used to discard a certain frequency and retain all higher and lower frequencies. Often a machine will exhibit one predominant vibration frequency; a notch filter can eliminate that frequency to observe the total machine action at all other frequencies.

Many common machine malfunctions have specific associated vibration frequencies. The use of filters to identify these frequencies would seemingly lead directly to the identification of particular machine malfunctions. There are even several charts published relating vibration frequency to different machine malfunctions.[5] It should be noted that frequency analysis charts alone are not the final answer to vibration analysis. The overall shape of the waveform and orbit as viewed on an oscilloscope and the measurement of phase angle are two very important vibration characteristics which are unavailable through frequency analysis. Determination of vibration frequency may even be the *first* step in machine problem solving, but it should not be the last.

Tunable Filter

A filter instrument which typically has several of the above functions and also an adjustment to control the tuned and cutoff frequencies is called a tunable filter. These instruments have a vibration readout meter so they can also be used for periodic overall measurements on unmonitored machines. A tunable filter may have two bandpass filters—broad and narrow. The broad filter will have a wide bandwidth or low Q, and vice versa for the narrow filter. A narrow filter may pinpoint a certain vibration frequency and associated amplitude more accurately, but it suffers from a lower response time. The narrow filter may be essential for definition of low-frequency components, but the broad filter will be more useful for higher frequencies and provides a better response time when tracking.

Most tunable filters have analog dc outputs for driving an XY plotter in order to generate amplitude-vs.-frequency spectrum information. Some even have an auto-sweep feature where the tuning section is swept automatically through a certain frequency range. Any filtered output is especially useful when viewed on an oscilloscope, particularly with a Keyphasor. The filtered signals of Fig. 6-2 were made by means of a tunable filter with the bandpass filter adjusted to running-speed frequency.

Vector Filter Phase Meter

The vector filter is a unique type of filter instrument since it displays vibration at a specific frequency as a true vector quantity. A given vibration frequency has both magnitude (amplitude) and direction (phase angle). The most common use of a vector filter is for observation of the unbalance response of a machine. The instrument requires inputs from two transducers, a Keyphasor and a vibration transducer. The bandpass filter in the instrument is automatically tuned to running-speed frequency by the Keyphasor; the vibration probe measures amplitude, and the two transducers together provide a measurement of vibration phase angle. These three measurements—shaft revolutions per minute, amplitude, and phase angle—have associated dc analog outputs for use with an XY plotter. As a machine runs from rest to operating speed, this information will produce a Bodé plot, Fig. 6-3. Since vibration due to rotor unbalance occurs at running-speed frequency, the Bodé plot represents the fundamental unbalance response of a

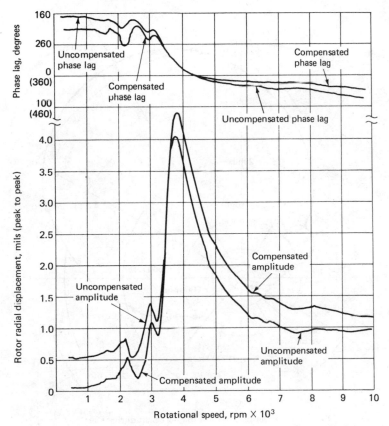

Figure 6-3 Bodé diagrams of compensated and uncompensated rotor displacement during start-up.

machine during start-up or shutdown. The Bodé plot is used to measure balance resonance frequencies (critical speeds), initial bow vectors, and amplification factors.[8]

Since an unbalance vector has magnitude and direction, it can be represented as a polar quantity. The vector filter will also generate two dc analog voltages for use with a plotter to represent these vector quantities on an XY basis. This information results in the polar plot, Fig. 6-4. The same information available from a Bodé plot is also provided by the polar plot. In addition, the polar representation shows the direction (phase angle) of the initial bow vector, the direction of initial rotor deflection due to unbalance, and the direction of rotor deflection in the self-balance speed region. Both the Bodé and polar plots are extremely useful in rotor balancing.

Spectrum Display

This instrument, often called a *spectrum analyzer,* consists of many bandpass filters, each fixed at a different frequency. The instrument usually provides a CRT display of vibration amplitudes (on the vertical scale) at specific frequencies (on the horizontal scale). The amplitude and frequency measurements are displayed continuously on the CRT and any changes in the spectrum content can be identified while the machine is running. These units usually incorporate an averaging function whereby transient data can be captured over several spectrum samples. A peak hold function allows the display

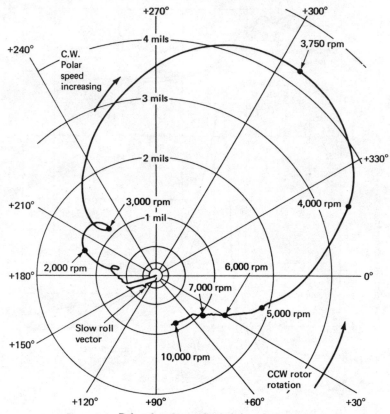

Figure 6-4   Polar plot of rotor displacement during start-up.

of the maximum amplitude at any frequency band over a period of time. Both amplitude and frequency measurements have associated dc analog outputs for use with an *XY* plotter.

Since many common machine malfunctions produce unique vibration frequencies, the spectrum display may be used to indicate the presence of these malfunctions and the relative magnitudes of each. A common use of this spectrum information is called *spectrum analysis* or *signature analysis*. This involves the comparison of vibration spectra taken at different points on a machine or the comparison of spectra taken at the same measurement location at different points in time. Figure 6-5 shows such a spectrum comparison. If spectra are acquired at different rotative speeds during a start-up or shutdown, the resulting presentation is the cascade or waterfall plot, Fig. 6-6.

Data Documentation

Previous sections have described the use of the oscilloscope camera and the plotter for hard-copy documentation of data presentations from various instruments. Recently, microprocessor and computer systems have been developed to display some of this same information on the computer CRT—oscilloscope waveforms and orbits, Bodé and polar plots, spectra, etc. Most computer CRTs are equipped with functions to provide direct hard-copy documentation via a graphics printer or outputs to a conventional plotter. One additional data-documentation instrument is used universally with all the above-men-

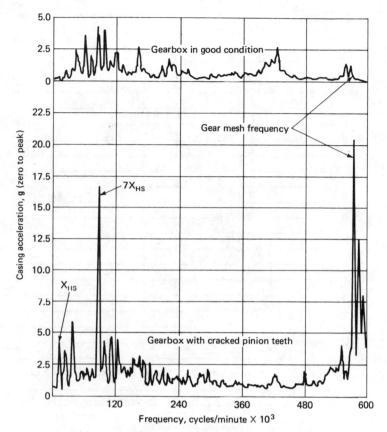

Figure 6-5 A comparison between gearbox casing acceleration characteristics.

tioned instruments—the magnetic-tape recorder. The tape recorder can capture dynamic and steady-state (in the case of FM recorders) transducer signals directly. Then in the playback mode, the signals can be sent from the recorder to these various data reduction instruments, exactly as they would have come from the transducers directly.

Tape recorders are especially useful when many measurement points need to be documented simultaneously, such as during a machine train start-up. Oscilloscopes, digital vector filters, and spectrum displays can of course be used directly with the transducers during the start-up, but most of these instruments are only one- or two-channel devices. However, after the start-up, the transducer signals can be reproduced from the magnetic tape and displayed and documented with each of these instruments independently.

INFORMATION SOURCES

Technical data and specification sheets for the various instruments used in machinery analysis are available from various instrument manufacturers. Most manufacturers also publish technical articles describing the specific uses of these products. Articles authored by persons in industry regarding machinery malfunction diagnosis are periodically published in industrial trade journals. Two of the best publications are *Hydrocarbon Processing* for the petrochemical industry and *Power* for the power generation industry. To

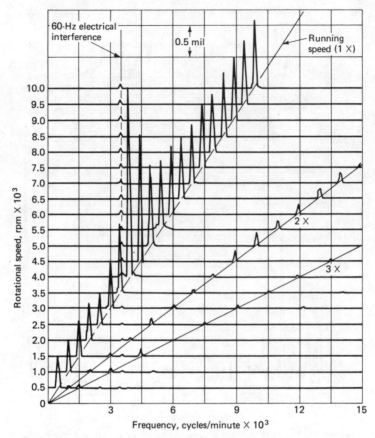

Figure 6-6 Cascade spectrum analysis of rotor displacement during start-up.

date, the most comprehensive book published on the subjects of machinery measurements, monitoring systems, data acquisition and reduction instruments, and machinery malfunction identification is by Charles Jackson, of Monsanto Company. This book, *The Practical Vibration Primer,* is published by Gulf Publishing Company, Houston, Texas. Other specific references are in the following list.

REFERENCES

1. Applications Note, "Vibration Measurement—Basic Parameters for Predictive Maintenance on Rotating Machinery," Bently Nevada Corporation, Minden, Nevada.
2. Dodd, V. Ray: "Machinery Monitoring Update," presented at the Texas A & M Sixth Turbomachinery Symposium, Houston, Texas., December 1977.
3. Standard 670, "Noncontacting Vibration and Axial Position Monitoring System," 1st ed., American Petroleum Institute, Washington, D.C., June 1976.
4. Applications Note, "Machinery Protection Systems for Various Types of Rotating Machinery," Bently Nevada Corporation, Minden, Nevada.
5. Jackson, Charles: *The Practical Vibration Primer,* 1979, Gulf Publishing Company, Houston, Texas.
6. Applications Note, "The Keyphasor—A Necessity for Machinery Diagnosis," Bently Nevada Corporation, Minden, Nevada.

7. Applications Notes, (1) "Alignment Loading of Gear Type Couplings," (2) "Preloads on Rotating Shafts," (3) "Shaft Position Changes Reveal Machinery Behavior/Malfunctions," (4) "Mechanical Degradation Due to Electrostatic Shaft Voltage Discharge," (5) "Attitude Angle and the Newton Dogleg Law of Rotating Machinery," and (6) "Plotting Average Shaft Centerline Position," Bently Nevada Corporation, Minden, Nevada.
8. Jackson, Charles: "Balance Rotors by Orbit Analysis," *Hydrocarbon Processing,* January 1971.
9. Standard 612, "Special-Purpose Steam Turbines for Refinery Services," 2d ed., American Petroleum Institute, Washington, D.C., June 1979.

section 2

Lubricants and Lubrication Systems

chapter 2-1

Lubricants: General Theory and Practice

by
Anne Bernhardt,
Staff Engineer
Gulf Research and Development Company
Petroleum Products Department
Pittsburgh, Pennsylvania

GENERAL THEORY

Functions of Lubricants

Lubricants perform a variety of functions. The primary, and most obvious, function is to reduce friction and wear in moving machinery. In addition, lubricants can

- Protect metal surfaces against rust and corrosion

- Control temperature and act as heat-transfer agents
- Flush out contaminants
- Transmit hydraulic power
- Absorb or damp shocks
- Form seals

Because reducing friction is such an important function of lubricants, it is necessary to understand how they perform.

Friction

Friction is the resistance to motion between two bodies in intimate contact. Two types of friction can be identified: solid (or dry) friction and fluid friction.

Solid Friction.   Solid friction occurs when there is physical contact between two solid bodies moving relative to each other. The type of motion divides solid friction into two categories, sliding and rolling friction.

Sliding Friction.   This is the resistance to movement as one body slides over another. Solid surfaces which appear smooth to the eye will in fact consist of many peaks and valleys. The resistance to motion is due primarily to the interlocking of these asperities. Under conditions of extreme pressure, the heat generated by sliding friction can result in welding of the points of contact.

Rolling Friction.   This is the resistance to motion as one solid body rolls over another. It is caused primarily by the deformation of the rolling elements and support surfaces under load. For a given load, rolling friction is significantly less than sliding friction.

Fluid Friction. Fluid friction occurs when two solid bodies in relative motion are completely separated by a fluid. It is caused by the resistance to motion between the molecules in the fluid. For a given load, fluid friction usually is significantly less than solid friction. The film thickness, relative to the height of the surface asperities, distinguishes three types of lubrication:

- Full or thick-film lubrication
- Mixed-film lubrication
- Boundary lubrication

Full or Thick-Film Lubrication. This exists when the lubricant film between two surfaces is of sufficient thickness to completely separate the asperities on the two surfaces. In this case, true fluid friction exists between the moving surfaces and no metal-to-metal contact will occur (Fig. 1-1a).

Mixed-Film Lubrication. This exists when the lubricant film between the two surfaces is of sufficient thickness to separate most of the surface asperities but some metal-to-metal contact may occur (Fig. 1-1b).

Boundary Lubrication. This exists when the film thickness is equal to the asperity heights and extensive metal-to-metal contact occurs (Fig. 1-1c).

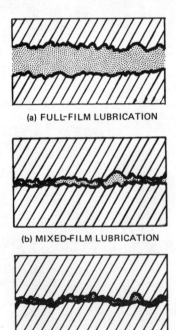

(a) FULL-FILM LUBRICATION

(b) MIXED-FILM LUBRICATION

(c) BOUNDARY LUBRICATION

Figure 1-1   (a) Full-film lubrication. (b) Mixed-film lubrication. (c) Boundary lubrication.

Formation of the Lubricant Film

The lubricant film may be formed and maintained in one of two ways:

- Hydrostatically
- Hydrodynamically

Hydrostatic Lubrication

Hydrostatic lubrication occurs when the film is formed by pumping the lubricant under pressure between the bearing surfaces. The surfaces may or may not be moving with respect to each other. The hydrostatic pressure acts to completely separate the surfaces, and full-film lubrication is established.

Hydrodynamic Lubrication

Hydrodynamic lubrication depends on motion between the two solid surfaces to generate and maintain the lubricating film. In a plain bearing that is not rotating, the shaft will rest on the bottom of the bearing and will tend to squeeze any lubricant out from between the surfaces. When the shaft begins to rotate, a very thin film of lubricant will tend to adhere to the shaft surface and will be drawn between the shaft and the bearing. A film that will ultimately separate the load-bearing surfaces is established. A lubricant film generated in this manner is called a *hydrodynamic film*.

The thickness of the hydrodynamic oil film developed in a properly designed plain bearing is dependent on the oil viscosity, the bearing load, speed, metallurgy, and quality of the bearing surfaces. The dimensionless bearing parameter ZN/P conveniently describes the combined effect of viscosity Z, speed N, and load P.

The thickness of the hydrodynamic film and the amount of friction developed in the bearing can be predicted by means of the bearing parameter ZN/P. Plotting the bearing coefficient of friction vs. the bearing parameter for a particular bearing and lubricant gives a characteristic curve similar to the one in Fig. 1-2. Experience has shown that the

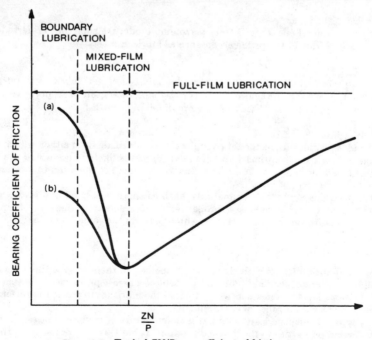

Figure 1-2 Typical ZN/P vs. coefficient of friction curve.

thickness of the lubricant film developed in a bearing can be determined by estimating where on the curve a bearing is operating (see Fig. 1-2).

Changes in the quality of the bearing's metallurgy or surface finish, or of the lubricant's "oiliness" or film strength, will cause shifts in coefficient of friction under boundary or mixed-film lubrication conditions. For example, when holding everything else constant, adding an oiliness additive to the lubricant will shift the bearing performance curve from curve *a* to curve *b*. As this indicates, it is possible to reduce the amount of friction generated in a particular bearing under boundary or mixed-film lubrication conditions through the use of certain additives.

TYPES OF LUBRICANTS

There are three major categories of lubricants:

- Fluid lubricants
- Greases (semisolid lubricants)
- Solid lubricants

Each lubricant will have its own physical properties which will affect its performance in different applications. A knowledge of the various types of lubricants on the market today and a basic understanding of their advantages and limitations is most helpful in the selection of the optimum lubricant for a particular application.

Fluid Lubricants

Fluid lubricants are the most widely used. The most common are petroleum oils, synthetic fluids, and animal or vegetable oils. Many other fluids can fulfill a lubrication function under special conditions when the use of oils may be precluded.

Petroleum or Mineral Oils

Petroleum or mineral oils, refined from petroleum crude, are sometimes referred to as *conventional oils* due to their wide acceptance as lubricants.

Synthetic Fluids

Synthetic fluids include all artificially made fluids used for lubricating purposes. Included in this category are synthesized hydrocarbons, esters, silicones, polyglycols, and phosphate esters. These are discussed in more detail in Chap. 2-2.

Animal and Vegetable Oils

Animal and vegetable oils, as the terms indicate, are oils made from either animal fat or vegetables. They are used primarily where food contact is likely to occur and the lubricant must be edible. Their main disadvantage is that most of them tend to deteriorate rapidly in the presence of heat.

In the past, oils made from animal fats, such as sperm whale oil and lard oil, were frequently used for their "oiliness" properties. Today, however, these are frequently being replaced with synthesized fatty oils which perform the same function.

Greases

Greases are fluid lubricants with thickeners dispersed in them to give them a solid or semisolid consistency. The fluid lubricant content of a grease performs the actual lubricating function. The thickener acts solely to hold the lubricant in place, to prevent leakage, and to block the entrance of contaminants.

Many types of thickeners are used in the manufacture of modern greases. Each type imparts certain properties to the finished product. Table 1-3 describes some of the typical properties and applications of greases manufactured with certain common thickeners.

Solid Lubricants

Solid lubricants, such as graphite, molybdenum disulfide (moly), and PFTE (*polytetrafluoroethylene*), are not only used by themselves but are also frequently added to oils and greases to improve their performance under boundary lubrication conditions. Chapter 2-3 discusses these in more detail.

IMPORTANT LUBRICANT CHARACTERISTICS

Various physical and chemical properties of lubricants are measured and used to determine a lubricant's suitability for different applications.

Oil Properties

Viscosity

Of the various lubricant properties and specifications, viscosity (also referred to as the "body" or "weight") normally is considered the most important. It is a measure of the force required to overcome fluid friction and allow an oil to flow.

Industry uses several different systems to express the viscosity of an oil. Table 1-1 gives a comparison of some of the most common. Lubricant specifications usually express viscosity in Saybolt Universal Seconds (SUS or SSU) at 100 and 210°F (37.8 and 98.9°C) and/or in centistokes (cSt) at 40 and 100°C (104 and 212°F). Viscosity expressed in centistokes is called the *kinematic viscosity*.

With the current move toward metrication and the establishment of the International Organization for Standardization (ISO) viscosity grade identification system, the centistoke has become the preferred unit of measure. The ISO viscosity grade system contains 18 grades covering a viscosity range from 2 to 1500 cSt at 40°C. Each grade is approximately 50 percent more viscous than the next lower grade.

Laboratories determine oil viscosity experimentally using a viscometer (Fig. 1-3). The viscometer measures an oil's kinematic viscosity by the time (in seconds) it takes a specified volume of lubricant to pass through a capillary of a specified size, at a specified temperature. The kinematic viscosity is then derived by calculations based on constants for the viscometer and the time it took the sample to pass through the instrument.

Viscosity Index

The viscosity index (VI) is an empirical measure of an oil's change in viscosity with temperature. The greater the value of the viscosity index, the less the oil viscosity will change with temperature. Originally ranging from 0 to 100, viscosity indexes greater than 100 are now achieved with certain synthetic oils or through the use of additives.

Oxidation Stability

When a lubricant is exposed to heat and air, a chemical reaction called *oxidation* takes place. Products of this reaction include carbonaceous deposits, sludge, varnish, resins, and corrosive and noncorrosive acids. Oxidation usually brings with it an increase in the viscosity of the oil.

The rate of oxidation is dependent on the chemical composition of the oil, the ambient temperature, the amount of surface area exposed to air, the length of time the lubricant has been in service, and the presence of contaminants which can act as catalysts to the oxidation reaction.

Depending on the intended end use of the oil, the oxidation stability will be measured or expressed in different ways. All of the oxidation stability tests are based on placing a

TABLE 1-1 Commonly Used Industrial Lubricant Viscosity Classifications

ISO/ASTM viscosity grade system			Former ASTM-ASLE grade no.†	AGMA lubricant no.‡	SAE viscosity no. (approximate)§	SAE gear viscosity no. (approximate)¶
ISO viscosity grade no. (ISO VG)	Kinematic viscosity, cSt at 40°C	Saybolt viscosity, SUS at 100°F (approximate)*				
2	1.98–2.42	32.8–34.4	32			
3	2.88–3.52	36.0–38.2	36			
5	4.14–5.06	40.4–43.5	40			
7	6.12–7.48	47.2–52.0	50			
10	9.00–11.0	57.6–65.3	60			
15	13.5–16.5	75.8–89.1	75			
22	19.8–24.2	104.6–126.0	105		5W	
32	28.8–35.2	149.1–181.7	150	1 (R & O)	10W	75W
46	41.4–50.6	214–262	215	2 (R & O, EP)	20W	
68	61.2–74.8	317–389	315	3 (R & O, EP)	20	80W
100	90.0–110	468–574	465	4 (R & O, EP)	30	
150	135–165	709–866	700	5 (R & O, EP)	40	85W
220	198–242	1047–1283	1000	6 (R & O, EP)	50	90
320	288–352	1533–1876	1500	7 (EP, comp)		
460	414–506	2214–2719	2150	8 (EP, comp)		140
680	612–748	3298–4044	3150	8A (EP, comp)		
1000	900–1100	4882–5994	4650			250
1500	1350–1650	7383–9060	7000			

*SUS viscosities are approximate and are based on typical 95 VI single-grade oils.
†Numbers are equivalent to the *Plant Engineering* magazine designation. The American Society of Lubrication Engineers (ASLE) and the American Society for Testing and Materials (ASTM) used a viscosity grade system based on SUS at 100°F. This system is now obsolete.
‡American Gear Manufacturers Association (AGMA) proposed revisions will equate AGMA lubricant no. to the ISO VG.
§Engine Oil Viscosity Classification—SAE J 300d.
¶Axle and Manual Transmission Lubricant Viscosity Classification—SAE J 306c.

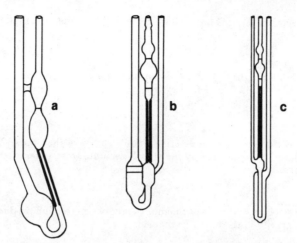

Figure 1-3 Capillary tube viscosimeters used to measure kinematic viscosity: (a) modified Ostwald, (b) Ubbelohde, (c) Fitz Simmons. (*Gulf Oil Corporation.*)

sample of oil under conditions which will greatly increase the rate of oxidation. Buildups of reaction products are then measured. The American Society for Testing and Materials (ASTM) D-943 test is the most widely used. Conducted under prescribed conditions, it measures the time (in hours) for the acidity of a sample of oil to increase a specified amount. The more stable the oil, the longer it will take for the change in acidity to occur.

Used-oil analysis to determine if the oil is suitable for further service is based on a comparison between the used oil and the new oil. Increases in viscosity, acidity, and buildups of insoluble contaminants are usually indicators that oxidation has occurred.

Thermal Stability

Thermal stability is a measure of an oil's ability to resist chemical change due to temperature. Since oxygen is present in most lubricant applications, the term *thermal stability* is frequently used in reference to the oxidation resistance of an oil.

Chemical Stability

Chemical stability defines an oil's ability to resist chemical change. Usually it, too, is used to refer to the oxidation stability of an oil. Chemical stability, other than resistance to oxidation, sometimes can refer to an oil's inertness in the presence of various metals and outside contaminants.

Carbon Residue

The carbon-forming tendencies of an oil can be determined with a test in which the weight percent of the carbon residue of a sample is measured after evaporation and pyrolysis.

Neutralization Number

The neutralization number (neut. no.) is a measure of the acidity or alkalinity of an oil. Usually reported as the total acid number (TAN) or total base number (TBN) it is expressed as the equivalent milligrams of potassium hydroxide required to neutralize the acidic or basic content of a 1-g sample of oil. Increases in the TAN or decreases in the TBN are usually indicators that oxidation has occurred.

Lubricity

Lubricity is the term used to describe an oil's "oiliness" or "slipperiness." If two oils of the same viscosity are used in the same application and one causes a greater reduction in friction than the other, it is said to have better lubricity than the first. This is strictly a descriptive term.

Saponification Number

The saponification number (SAP no.) is an indicator of the amount of fatty material present in an oil. The SAP no. will vary from 0, for an oil containing no fatty material, to 200 for 100 percent fatty material.

Demulsibility

Demulsibility is the term used to describe an oil's ability to shed water. The better the oil's demulsibility, the more rapidly the oil will separate from water after the two have been mixed together.

API Gravity

API gravity is a relative measure of the unit weight of a petroleum product. It is related to the specific gravity in the following manner:

$$\text{API gravity} = \frac{141.5}{\text{specific gravity}} - 131.5$$

Pour Point

Pour point is the lowest temperature at which an oil will flow in a certain test procedure. It is usually not advisable to use an oil at temperatures lower than 15°F (8°C) above its pour point.

Flash Point

Flash point is the oil temperature at which vapors from the oil ignite when an open flame is passed over a test sample.

Fire Point

Fire point is the oil temperature at which vapors from the oil will sustain a continuous flame. The fire point is usually approximately 60°F (33°C) above the flash point.

Grease Properties

Penetration

Penetration is an indicator of a grease's relative hardness or softness and not a criterion of quality. Measured on a penetrometer at 77°F (25°C), it is the depth of penetration (in tenths of millimeters) into the grease of a standard 150-g cone. The softer the grease, the greater the penetration number will be.

If the penetration test is performed on an "undisturbed" sample, the results are reported as unworked penetration. If the sample has been subjected to extrusion by a reciprocating perforated piston for a number of strokes (most commonly 60 strokes) prior to the penetration test, the results are reported as worked penetration. It is normally desirable to have as little difference between the worked and unworked penetrations as possible.

NLGI Consistency Numbers

The National Lubricating Grease Institute (NLGI) has developed a number system ranging from 000 (triple zero) to 6 to identify various grease consistencies. This system is used by most of industry. Table 1-2 gives the NLGI numbers, their corresponding worked penetration ranges, and their descriptions (their corresponding consistencies). Most multipurpose greases are of either a no. 1 or no. 2 consistency.

TABLE 1-2   NLGI Classification of Greases

NLGI consistency grade	Worked penetration ASTM D 217-60T	Description
000	445–475	Very fluid
00	400–430	Fluid
0	355–385	Semifluid
1	310–340	Very soft
2	265–295	Soft
3	220–250	Semistiff
4	175–205	Stiff
5	130–160	Very stiff
6	85–115	Hard

Dropping Point

Dropping point is the temperature at which a grease liquefies and will flow. Generally it is not advisable to use a grease at temperatures higher than 50°F (28°C) below its dropping point.

Soap

The thickener used to manufacture greases can be called "soap." Many greases use metallic soaps as thickeners. Table 1-3 shows a comparison of some of the key properties of greases manufactured with different soaps and their typical applications.

Fillers

Solid lubricants, particularly molybdenum disulfide, are frequently added to greases to enhance their performance. These solid lubricants are called *fillers*.

ADDITIVES USED IN LUBRICATING OILS

It is possible, through the use of chemical additives, to improve a lubricant's natural ability to protect metal surfaces, to resist chemical changes, and to drop out contaminants.

Since industrial lubricating oils are frequently described by the additives they contain, it is helpful to understand the function of the major types of additives. Following are general definitions of some of the most common, listed in alphabetical order:

Air release agents   Assist the oil in the release of entrapped air.

Antifoam agents   Promote the rapid breakup of foam bubbles.

Antiseptic agents or bactericides   Prevent the growth of microorganisms and bacteria. These are found primarily in water-soluble oils.

Antiwear agents   Decrease the coefficient of friction and reduce wear under boundary or mixed-film lubrication conditions.

Demulsifiers   Assist the natural ability of an oil to separate rapidly from water. These agents can be helpful in preventing rust since they help to keep water out of the oil and thus away from the metal surfaces.

Detergent-dispersant agents   Prevent the formation of varnish and sludge. They are most commonly found in engine oils.

Emulsifiers   Permit the mixing of oil and water to form stable emulsions. They are used primarily in the manufacture of water-soluble oils.

TABLE 1-3 Grease Application Guide

				Thickener			
	Lithium	Calcium	Sodium	Calcium complex	Aluminum complex	Polyurea	Clay or "Bentone"
Properties							
Dropping point, °F	350–375	200–225	325–350	500+	500+	550	
Average max usable temp, °F	275	175	250	325	300	350	275
High-temperature characteristics	Good	Poor	Fair-Good	Good	Good	Excellent	Good
Thermal stability	Good	Poor	Fair-Good	Good	Good	Excellent	Good
Low-temperature characteristics	Good	Fair	Fair	Fair–good	Fair–good	Good	Good
Pumpability	Excellent	Fair–good	Poor	Fair	Fair–good	Good–excellent	Good
Mechanical stability	Excellent	Good	Poor	Fair–good	Excellent	Excellent	Good
Oil separation	Good	Poor	Fair–good	Excellent	Good	Excellent	Excellent
Water resistance	Good	Excellent	Poor	Good	Excellent	Excellent	Good
Texture	Smooth and buttery	Smooth and buttery	Buttery to fibrous	Smooth and buttery	Smooth and buttery	Smooth and buttery	Smooth and buttery
Rust protection	Fair to good	Poor	Excellent	Excellent	Good	Excellent	Poor
Oxidation stability	Fair to good	Poor	Poor	Good	Good	Excellent	Good
Other properties			Good adhesive and cohesive properties	Good inherent EP properties			
Applications							
	Multipurpose All applications except extra high temperatures	Where water is dominant factor Wet, moderate low temperature conditions Plain and roller bearings, water pumps, slides	Antifriction and plain bearings Electric motors, fans Must be used in dry conditions	High temperatures Corrosive conditions Do not use in centralized lubrication systems	Multipurpose Moderately high temperatures	Multipurpose High temperatures Antifriction and plain bearings Electric motors, fans Wet conditions Corrosive conditions	Multipurpose High temperatures

Extreme-pressure agents   Protect against metal-to-metal contact and welding after the oil film has been ruptured by high loads or sliding velocities. The majority of the extreme-pressure oils on the market today are of the sulfur-phosphorus type and are noncorrosive to most metals including brass. This was not true of some of the earlier formulations, and many misconceptions still exist in this regard.

"Oiliness" or fatty compounds   Improve the lubricity or slipperiness of an oil. These compounds are also helpful in resisting water wash-off.

Oxidation inhibitors   Prevent or retard the oxidation of an oil, thereby reducing the formation of deposits and acids.

Pour-point depressants   Lower the pour point of paraffinic petroleum oils.

Rust and corrosion inhibitors   Improve an oil's ability to protect metal surfaces from rust and corrosion. ·

Tackiness agents   Improve the adhesive qualities of an oil.

Viscosity index improvers   Increase the viscosity index of an oil by increasing an oil's viscosity at high temperatures. These additives are most widely used in motor oils to create multigrade oils.

LUBRICANT SELECTION

Practical lubrication is more an art than an exact science. Proper lubricant selection depends on the equipment design, the operating conditions, and the method of application.

Most equipment manufacturers provide lubrication recommendations based on design, normal operating conditions, and past experience. Whenever possible, these recommendations should be followed. In addition, most reputable oil suppliers keep in close contact with major equipment builders and are available to consult with users on lubricant selection.

The recommendations included in this chapter are based on standard practices and are intended solely as guidelines.

General Selection Guidelines

The design of the equipment and the expected operating conditions will determine which functions the lubricant is expected to perform and will dictate the type of lubricant and additives that will be best suited.

The oil of proper viscosity for an application is a function of speed, load, and ambient temperature. Conditions of high loads and slow speeds will require a high-viscosity oil. Similarly, a low-viscosity oil is best suited to conditions of low loads and high speeds. Ideally, one would like to select the oil of lowest possible viscosity that is capable of maintaining a lubricant film between the moving surfaces. Selection of a higher-viscosity oil than is needed can result in power losses and temperature buildups due to the higher internal fluid friction of the lubricant.

The effect of operating temperatures on the selection of the lubricant should not be overlooked. Since oil viscosity decreases as temperatures increase, it is necessary to select higher-viscosity fluids for high-temperature applications and lower-viscosity fluids for low-temperature applications in order to ensure adequate lubricant film thickness and minimal fluid friction. Fluids with high viscosity indexes (high VI) should be used for applications where wide temperature ranges are anticipated.

Operating Limits of Petroleum Oils

As a result of additive technology, a suitable petroleum-base lubricating oil can be found for most applications. Exceptions can exist where fire-resistant fluids are required or extreme temperature conditions exist.

Where fire-resistant fluids are required, petroleum oils are not suitable. In some instances water-oil emulsions are acceptable if operating temperatures are below 150°F (65°C) (to avoid excessive water evaporation) and if the equipment is designed to handle these fluids. Two types of water-oil emulsions are currently in use. The high-water-base fluids, (95/5 fluids as they are sometimes called). are an oil-in-water emulsion containing 95% water and 5% oil. The other is the invert emulsion, a water-in-oil emulsion, which contains approximately 40% water.

Unsatisfactory performance of petroleum oils can occur under three types of extreme temperature conditions:

- Excessively high temperatures
- Excessively low temperatures
- Wide temperature variations

Petroleum oils will adequately withstand very high temperatures for very short periods of time. Problems will occur when the oil is subjected to high temperatures for extended periods of time. The rate of oxidation of petroleum oils subjected to constant temperatures above 115°F (45°C) will approximately double for every 15 to 20°F (8 to 10°C) rise in temperature. Temperatures above 200°F (95°C) will almost always result in excessive sludge and deposit formation and should be avoided. In circulating systems, the reservoir should always be cool enough to comfortably hold a hand on it. The oxidation rate is usually negligible at temperatures below 115°F (45°C).

Petroleum oils should not be used at temperatures less than 10 to 15°F (5 to 8°C) above their pour point. For applications subjected to large temperature variation only high-viscosity-index fluids should be used. The viscosity index should be high enough to ensure that the oil viscosity remains within the recommended limits at both the high- and low-temperature extremes to which the equipment is subjected.

Plain-Bearing Lubrication

Plain bearings, also called journal or sleeve bearings, comprise one of the simplest machine components. The type of motion between the bearing and the shaft is pure sliding.

In plain bearings, the lubricant must reduce sliding friction, carry away any heat generated in the bearing, prevent rust and corrosion, and serve as a seal to prevent the entry of foreign material.

Barring any unusual operating conditions, plain bearings will operate satisfactorily with any lubricant of the correct viscosity. Special operating conditions may require the use of oils containing additives. Antiwear and extreme-pressure oils may be desirable for plain bearings operating intermittently or under very high loads. Rust- and corrosion-inhibited oils are generally preferred for humid operating environments.

Most plain bearings are designed to operate under full-film hydrodynamic lubrication. Referring to Fig. 1-2, and assuming the bearing load and oil viscosity to be constant, the lubricant film development would be expected to follow the ZN/P curve as the shaft speed increases. If oil of the proper viscosity is selected for the load and speed conditions, full-film hydrodynamic lubrication will prevail during continuous operation.

Numerous mathematical models of plain-bearing lubrication have been used in attempts to accurately select the best oil viscosity for a plain bearing. Unfortunately, these models are complicated and expensive to develop. For this reason, except in special cases, lubricant viscosity selection is usually based on standard practices established through experience. Table 1-4 presents a general guide for viscosity selection for plain bearings subjected to average loading.

Plain bearings may be grease-lubricated if their operating speed does not exceed approximately 6 ft/s (2 m/s). At higher speeds, excessive temperature buildup could result.

In general, relatively soft greases are used for centralized systems and harder greases

TABLE 1-4 Oil Viscosity Selection for Plain Bearings

Bearing speed factor, r/min × shaft diameter		Viscosity at operating temperature	
in	mm	cSt	SUS
Below 750	Below 1,900	130–325	600–1500
750–2,000	1,900–50,800	65–130	300–600
2,000–4,000	50,800–101,600	32–65	150–300
4,000–10,000	101,600–254,000	14–32	75–150
Above 10,000	Above 254,000	5–14	40–75

for compression cups and open journals. Each application should be considered on its own merits, taking into consideration the operating conditions. Temperature and water contamination require particular attention.

Plain bearings are frequently grooved (Fig. 1-4) to improve the distribution and flow of the lubricant. Normally, two important rules should be followed when grooving a plain bearing:

- Grooves should not extend into the load-carrying area of the bearing because this would increase unit pressures.
- Groove edges should be rounded to prevent scraping the lubricant off the journal.

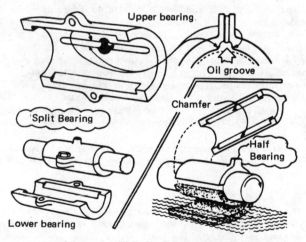

Figure 1-4 Oil groove. (*U.S. Steel Corporation, Reprinted from* The Lubrication Engineers Manual, *Copyright 1971.*)

Antifriction-Bearing Lubrication

Antifriction or roller bearings use balls or rollers to substitute rolling friction for sliding friction. This type of bearing has closer tolerances than do plain bearings and is used where precision and higher speeds are required.

In antifriction bearings a lubricant facilitates easy rolling, reduces the friction generated between the rolling elements and the cages or retainers, prevents rust and corrosion, and serves as a seal to prevent the entry of foreign material.

High-quality rust- and oxidation-inhibited (R & O) oils are generally recommended, especially where high-temperature conditions may oxidize the oil and so lead to the formation of deposits which could interfere with the free action of the rolling elements.

TABLE 1-5 Oil Viscosity Selection for Antifriction Bearings

Speed factor bearing bore, mm × r/min	Operating temperatures		Viscosity	
	°F	°C	ISO viscosity grade	SUS at 100°F
Up to 75,000	−40–32	−40–0	15–32	70–150
	32–150	0–65	32–100	150–600
	150–200	65–93	100–220	600–1200
	200–250	93–21	220–680	1100–3000
75,000–200,000	−40–32	−40–0	7–22	50–100
	32–150	0–65	22–68	100–300
	150–200	65–93	68–100	300–600
	200–250	95–121	150–320	700–2100
200,000–400,000	−40–32	−40–0	7–15	50–70
	32–150	0–65	15–46	70–200
	150–200	65–93	32–68	150–300
	200–250	93–121	68–150	400–900
Above 400,000	−40–32	−40–0	5–10	40–60
	32–150	0–65	10–32	60–150
	150–200	65–93	22–46	100–200
	200–250	93–121	68–100	300–600

Extreme pressure and antiwear additives may also be desirable under conditions of heavy or high shock loads.

Because of its better cooling ability, oil is generally preferred to grease. Table 1-5 gives general guidelines to the proper viscosity selection of oils for antifriction bearings.

Grease may be used to lubricate antifriction bearings running at low speeds and located in areas where they are likely to receive infrequent attention.

The selection of the proper type and grade of grease depends on the operating conditions and the method of application. Generally, soft greases (i.e., NLGI no. 1 consistency) with low base oil viscosity are preferred for use at low temperatures and in central systems. Harder greases (i.e., NLGI no. 2 consistency) with low base oil viscosity perform better at high speeds.

Care should be taken not to overgrease antifriction bearings because this can lead to excessive temperature buildup. Generally the bearing housing should be ⅓ to ½ filled.

Gear Lubrication

The motion between gear teeth as they go through mesh is a combination of sliding and rolling. The type of gear, the operating load, speed, temperatures, method of application of the lubricant, and metallurgy of the gears, are all important considerations in the selection of a lubricant.

Industrial gearing may either be enclosed, in which case the gears and the bearings which support them are operated off the same lubricant system; or open, in which case the mountings are lubricated separately from the gears themselves.

Due to the high sliding forces encountered in enclosed worm and hypoid gears, lubricant selection for these should be considered separately from lubrication of other types of enclosed gears.

As with all equipment, the first rule in selecting a gear lubricant is to follow the manufacturer's recommendation, if at all possible. In general, one of the following types of oils is used:

Rust- and Oxidation-Inhibited Oils. R & O oils are good-quality petroleum-based oils containing rust and oxidation inhibitors. These oils provide satisfactory protection for most lightly to moderately loaded enclosed gears.

Extreme-Pressure Oils. EP oils are usually high-quality petroleum-based oils containing extreme-pressure additives. These products are especially helpful when high-load conditions exist and are a must in the lubrication of enclosed hypoid gears.

Compounded Oils. These are usually petroleum-based oils containing 3 to 5 percent fatty or synthetic fatty oils (usually animal fat or acidless tallow). They are usually used for worm-gear lubrication where the fatty content helps reduce the friction generated under high sliding conditions.

Heavy Open-Gear Compounds. These are very-heavy-bodied tarlike substances designed to stick to the metal surfaces. Some are so thick they must be heated or diluted with a solvent to soften them for application. These products are used in cases where the lubricant application is intermittent.

A large number of gear lubrication models and viscosity selection guides exist. In the United States, the most widely used selection method employs the American Gear Manufacturers Association (AGMA) standards. Under its specifications for enclosed industrial gear drives the AGMA has defined lubricant numbers which designate viscosity grades for gear oils. These grades are currently being revised to correspond to the ISO viscosity grades. Table 1-1 identifies the newly proposed AGMA viscosity numbers with their corresponding ISO viscosity grades.

As a rule, low speeds and high pressures require high-viscosity oils. Intermediate speeds and pressures require medium-viscosity oils, and high speeds and low pressures require low-viscosity oils. Table 1-6 gives some very broad guidelines for viscosity and type of lubricant for industrial gearing.

Open gears operate under conditions of boundary lubrication. The lubricant can be applied by hand or via drop-feed cups, mechanical force-feed lubricators, or sprays.

Heavy-bodied oils with good adhesive and film-strength properties are required because centrifugal forces tend to throw the lubricant off the gear teeth.

TABLE 1-6   Oil Selection for Enclosed Gear Drives

Service	ISO viscosity grade	Oil type
Helical, herringbone, straight-bevel, spiral-bevel, and spur-gear drives		
Operating at normal speeds and loads	220	EP or R & O
Operating at normal speeds and high loads	220	EP
Operating at high speeds (above 3600 r/min)	68	EP or R & O
Worm-gear drives	460	Compounded or EP
Hypoid-gear drives		
Normal speeds (1200–2000 r/min)	220	EP
High speeds (above 2000 r/min)	150	EP
Low speeds (below 1200 r/min)	460	EP

Compressor Lubrication

The compressor model and type, the loading, the gas being compressed, and other environmental conditions dictate the type and viscosity of the oil to be used. Most compressors are lubricated with petroleum oils; however, there has been considerable interest in synthetic lubricants for compressor lubrication in recent years.

Compressing gases other than air creates problems which require special lubrication consideration because of possible chemical reactions between the gas being compressed and the lubricant. Since no two cases are alike, it is recommended that the compressor manufacturer and lubricant supplier be consulted for recommendations for a particular operation.

Oils for use in compressors should have the following characteristics:

Good Stability

A good compressor oil must have high oxidation stability to minimize the formation of gum and carbon deposits. Such deposits can cause valve sticking. This can lead to very-high-temperature conditions and compressor malfunction.

Good Demulsibility

A good compressor oil must be able to shed water readily to prevent formation of emulsions which could interfere with proper lubrication.

Anticorrosion and Antirust Properties

A compressor lubricant must protect the valves, pistons, rings, and bearings against rust and corrosion. This is especially important in humid atmospheres or in compressors that operate intermittently.

Good Antiwear Properties

Good compressor oils must form and maintain a strong lubricant film at relatively high temperatures; therefore, good antiwear properties are required.

Nonfoaming Properties

This requirement is especially important in crankcases where air-oil mixtures could impair good lubrication.

Low Pour Point

This property is necessary only for low-temperature start-up. Usually it is a factor only in portable air compressors which will frequently be used outdoors.

Proper Viscosity

Table 1-7 summarizes the oil viscosity requirements for various types of compressors. The operator's manual should be consulted for the manufacturer's viscosity recommendations for the prevailing operating temperatures and conditions.

TABLE 1-7   Lubricant Selection for Compressors

Type of compressor	Type of service	Recommended ISO viscosity grade
Reciprocating		
Crankcase	All	68–100
Cylinders		
Under 300 lb/in^2	Dry air	68–100
	Wet air	100
Over 300 lb/in^2	Dry air	100–150
	Wet air	220
Rotary		
Sliding-vane type		
Air- or water-cooled	Dry air	32–68
	Wet air	46–68
Oil-flooded	Dry air	150
	Wet air	150
Lobe or impeller type	Air	32–46
Liquid-piston type	Air	32
Dynamic compressors		
Centrifugal	Air	32
Axial-flow	Air	32

SPECIFICATIONS AND STANDARDS

"Lubrication of Industrial Enclosed Gear Drives," AGMA Standard 250.04 American Gear Manufacturers Association, Arlington, Va., November, 1974.

"Lubrication of Industrial Open Gearing," AGMA Standard 251.02 American Gear Manufacturers Association, Arlington, Va., November 1974.

"ASLE Standards for Machine Tool Petroleum Fluids," American Society of Lubrication Engineers, Park Ridge, Ill., January 1972.

Annual Book of ASTM Standards, Parts 23, 24, 25, and 47; American Society for Testing and Materials, Philadelphia.

BIBLIOGRAPHY

Bailey, Charles A., and Joseph S. Aarons (eds.): *The Lubrication Engineers Manual,* United States Steel Corp., Pittsburgh, 1971.

Bearings and Their Lubrication, Gulf Oil Corp., Pittsburgh, 1952.

Billett, Michael: *Industrial Lubrication: A Practical Handbook for Lubrication and Production Engineers,* Pergamon, Oxford, England, 1979.

Brewer, Allen F.: *Basic Lubrication Practice,* Reinhold, New York, 1955.

Brewer, Allen F.: *Effective Lubrication,* Robert E. Krieger, Huntington, N.Y., 1974.

Ellis, E. G.: *Fundamentals of Lubrication,* 2d ed., Scientific Publications, Broseley, England, 1970.

Fuller, Dudley D.: *Theory and Practice of Lubrication for Engineers,* Wiley, New York, 1956.

Gunther, Raymond C.: *Lubrication,* Chilton, Philadelphia, 1971.

Neale, M. J. (ed.): *Tribology Handbook,* Wiley, New York, 1973.

O'Connor, J. J. and J. Boyd (ed.): *Standard Handbook of Lubrication Engineers,* McGraw-Hill, New York, 1968.

Shigley, Joseph E.: *Mechanical Engineering Design,* 2d ed., McGraw-Hill, New York, 1972.

chapter 2-2

Synthetic Lubricants

by
Marvin Campen,
Product Manager—Synthetic Fluids
Gulf Oil Chemicals Company
Houston, Texas

DEFINITION AND CLASSIFICATION

Synthetic lubricants, or synlubes, are made by compounding (or blending) chemically synthesized base fluids with conventional lubricant additives. (For some applications the synthesized base fluids are used neat, without additives.) Synthesized base fluids are formed by combining low-molecular-weight (MW) components via chemical reactions into higher-MW compounds. Thus each base fluid's molecular structure is planned and controlled and its properties are predictable. Synthetic base fluids do not occur in nature, as does the crude petroleum oil from which conventional mineral-oil-base stocks are derived by refining processes. Only by *super-refining,* which has never been economically feasible, could petroleum-base stocks be obtained with properties comparable to those of synthesized base fluids. Yet most synthetic base fluids come from petroleum via synthesis using components derived from petroleum hydrocarbons.

Principal reasons for the increasing uses of synlubes are their abilities to:

- Work where conventional lubricants will not
- Comply with certain specifications or regulations (as military, OSHA, safety, pollution)
- Provide enhanced cost effectiveness, including energy savings.

The distinguishing feature of all synthetics is superiority in one or more respects to the mineral oils. Advantageous characteristics of the most widely used synlubes, in varying degrees, are:

- Low-temperature fluidity
- High-temperature oxidation stability and fire resistance (high flash, fire, and autoignition points)
- Low volatility in relation to viscosity
- High viscosity index (Less change of viscosity with temperature)

Some synthetics are chemically inert; some are fire-resistant or nonflammable.

Thus synlubes are high-performance, problem-solving lubricants that provide operating benefits such as: (1) less frequent lubrication ("sealed for life" in some machines); (2) less maintenance; (3) higher productivity; (4) lower parts rejection (due to more uniform tolerances and quality); (5) longer machine life; (6) reduced fire hazard; (7) greater resistance to acids, bases, and solvents; (8) more economical metalworking-oil or coolant applications; and (9) easier reclamation or disposal.

Although synthetic base fluids make possible these benefits, proper compounding of base fluids with performance additives is essential to achieving finished synlubes with the desired advantages. Getting one's money's worth from a more-expensive, high-quality synlube is best assured by buying specification-grade, approved synlubes from reputable suppliers who know how to maximize the advantages of each base fluid's properties which include (1) *rheology,* or flow characteristics; (2) *volatility;* (3) *stability,* thermal, oxidative, hydrolytic, chemical; (4) *additives' compatibility* and *response;* and (5) *effects on elastomers, paint, and other softwear.*

The *classification* of synthetic base fluids was proposed by an ad hoc ASTM committee and subsequently adopted by SAE in Standard J 357a and by RCCC, as follows:

1. **Synthetic Hydrocarbons**
   Alkylated aromatics (dialkylbenzenes, DAB) and cycloaliphatics
   Polyalphaolefins (PAO)
   Polybutenes
2. **Organic Esters**
   Dibasic acid esters (diesters)
   Polyol esters
   Polyesters
3. **Other**
   Halogenated hydrocarbons
   Phosphate esters
   Polyglycols (polyalkylene glycols, PAG)
   Polyphenyl ethers
   Silicate esters
   Silicones
4. **Blends**   Synthetic base stock may be mixtures of the above which are compatible.

By specific omission from this classification, other compounds and mixtures such as the following are *not classified* as synthetic lubricants: (1) molybdenum disulfide (MoS_2), graphite, and other solids; (2) plastics (PE, PTFE, nylon, et al.); (3) inert and liquefied gases; (4) liquid metals (as Na, K); (5) exotic natural products (as jojoba oil); and (6) surfactant-chemical-concentrate, high-water-content, or "water-additive" coolants and other metalworking fluids (which are nevertheless called "synthetic" by some people in the metalworking industry).

Unless a lubricant consists of at least 90 percent synthetic base fluid, it should not be designated a synlube; rather it should be labeled "partial synthetic," "fortified synthetic," "semisynthetic," or some similar term.

PROPERTIES AND USES OF SYNTHETIC LUBRICANTS

The most widely used industrial synlube classes are: (1) polyalphaolefins, (2) organic esters, (3) phosphate esters, (4) polyalkylene glycols, and (5) silicones. Their operating-temperature ranges are shown in Fig. 2-1, their performance vs. mineral oil, in Table

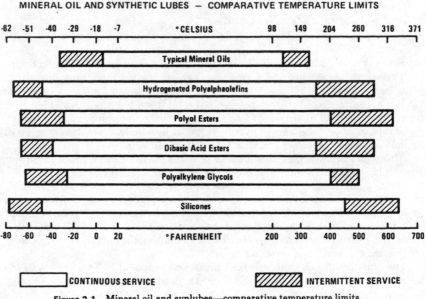

Figure 2-1   Mineral oil and synlubes—comparative temperature limits.

2-1. Most are supplied in several viscosity grades and can be formulated with proper additives to make industrial lubricants—circulating oils, gear and bearing lubes, and greases and other functional fluids for use with gears, traction drives, compressors, pumps, turbines, calenders, motors, hydraulic systems, valves, instrumentation, and other machinery and equipment, including metalworking applications (see Table 2-2). Many companies compound such lubricants from their own or from purchased synthetic base fluids and additives. Viscosity grades range from 1.5 cSt at 212°F (100°C) to firm greases.

Polyalphaolefins

Polyalphaolefins (PAO), or olefin alpha oligomers, are derived from linear alphaolefins which are made from ethylene, a basic building block from petroleum or natural gas liquids. Manufacturers are Bray, Ethyl/Cooper, Gulf, Lubrizol, Mobil, Emery, and Uniroyal. Shell, Conoco, Texaco, Arco, and others are potential manufacturers.

The largest volume use for PAO is in crankcase oils for internal combustion engines—gasoline, diesel, and natural gas. Principal industrial PAO lubricants are for gears, calenders, textile machinery, conveyors, gas turbines, hydraulic systems, instrumentation, rotary-screw and reciprocating compressors, pumps, equipment, and many grease applications—particularly in harsh environments (chlorine, HCl, and hot process gases). PAO products serve well as lubricants and sealant fluids for rotary mechanical seals of chemical process pumps and agitated kettles. The longer functional life of a PAO grease in a

TABLE 2-1　Relative Performance of Synlubes vs. Mineral Oil*

Properties	Mineral oil	Synthesized hydrocarbons		Organic esters		Polyalkylene glycol (PAG)	Phosphate ester	Silicone fluid
		Polyalpha-olefin (PAO)	Dialkylated (C_{12}) benzene (DAB)	Dibasic	Hindered polyol			
Viscosity-temperature properties (VI)	F	G	F	VG	G	G	P	E
Low-temperature fluidity, low pour point	P	G	G	G	G	G	F	G
High-temperature oxidation resistance with inhibitors	F	VG	G	G	E	F	F	G
Compatibility with mineral oils	E	E	E	G	F	P	F	P
Low volatility	F	E	G	E	E	G	G	G
Effect on most paints and finishes	N	N	N	S	M	M	C	S
Stability in presence of water (hydrolytic stability)	E	E	E	F	F	VG	F	G
Antirust properties, with inhibitor	E	E	E	F	F	G	F	G
Additive solubility	E	G	E	G	G	F	G	P
Elastomer swelling tendency—buna rubber	L	N	L	M	H	L	H	L

*Letter signifies performance level: P = poor, F = fair, G = good, VG = very good, E = excellent, M = moderate, H = high, C = considerable, N = none or nil, S = slight, L = light.

TABLE 2-2   Principle Uses of Synthetic Lubricants*

	Int. comb. engines	Gears and bearings	Greases	Compressors turbines†	Hydraulic fluids†	Water emulsions
Polyalphaolefins	E	E	E	VG	E	F
Organic esters	G	VG	VG	VG	VG	P
Polyalkylene glycols	——	VG	G	G	E(FR)	E
Phosphate esters	——	——	——	VG(FR)	E(FR)	——
Silicones	——	VG	E	E	E	——
Mineral oils	VG	VG	VG	VG	VG	F

*Letter signifies performance level: P = poor, F = fair, G = good, VG = very good, E = excellent
†FR = Fire-resistant.

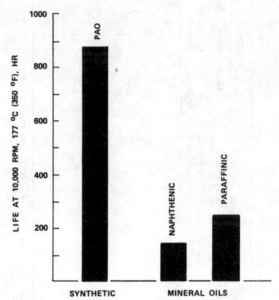

Figure 2-2   Functional-life ball-bearing test for grease, Federal Standard 791, No. 333.

high-speed ball-bearing test is shown in Fig. 2-2. Food-grade synlubes, compounded from FDA-approved PAO white oils, also are available.

Because PAO lubes are compatible with mineral oils and with existing systems designed for mineral oil products, they can be used in such systems without changing seals, hoses, or materials of pumps and other equipment. PAOs are often incorporated into organic ester lubes to assure seal compatibility.

Organic Esters

Diesters

Diester, or dibasic acid ester, base fluids are made by reacting short-chain C_{8-13} oxo alcohols with dibasic acids, such as adipic and phthalic (from petroleum), azelaic (from tallow), and dimer acids (from tall oil). Diesters are a subclass of polyesters, which include tri- and tetra-esters. More than 50 United States companies have the capacity for manufacturing at least 2×10^8 gal/year of these and many other diesters, primarily for plasticizers for PVC (vinyl) products. Manufacturers specializing in lubricant-grade diesters are Emery, Exxon, Hatco, Inolex, Mobil, Röhm and Haas, Tenneco, and Witco.

Diester lubricants are used mainly in oil-flooded air and natural-gas compressors. Their polar nature, due to the carbonyl group in their chemical structure, facilitates additive acceptance and provides solvation properties that help maintain clean intake and exhaust valves, thereby extending lubricant drain periods and improving ignition safety. Several makers of reciprocating compressors specify diester lubricants for some models. Diester lubricants are used in oven chain lubricants, in some gas-turbine engine oils, and as greases in certain low-temperature applications.

Polyol Esters

Polyol, or neopentyl, esters are formed by combining, linear short-chain (C_{5-10}) fatty acids with polyols, usually pentaerythritol (PE) or trimethylolpropane (TMP). Such "hindered esters" exhibit very high thermal stability, inherently higher than diesters. Emery, Hatco, Hercules, Witco/Humko, Mobil, and Stauffer are the current manufacturers of lubricant-grade polyol esters.

Polyol esters are chiefly used in military and commercial jet aircraft and in surface gas-turbine engines supplied by the jet-engine makers (Pratt & Whitney—TP & M, Detroit Diesel-Allison, GE aircraft-type models). Excellent gear oils, greases, and other industrial lubricants can be formulated from polyol ester fluids. However, their use in industry has been constrained by comparably performing, but lower-cost, diester and PAO lubricants. Polyol esters are sometimes combined with PAO to enhance solubility of certain additives and to optimize elastomer seal swell characteristics.

Polyalkylene Glycols

Polyalkylene glycols (PAG or polyglycols) are synthesized by combining ethylene oxide and propylene oxide under a variety of processing conditions. Consequently PAGs are available in the widest range of viscosities and hydrophobic/hydrophilic balances of any synthetic functional fluid (Table 2-3). The fluids are noncarbonizing and possess high viscosity indices. A limiting characteristic is their incompatibility with mineral oils and several conventional lubricant additives. Union Carbide is the dominant supplier. BASF, Dow, Olin, and Texaco also supply many PAG grades.

A large use for PAG is in metalworking fluids, e.g., coolants. The different degrees of water solubility of the different grades enable metal heat-treating quenchants to be selected for the desired rate of heat removal. Other large-volume applications of compounded PAG include heat-transfer oils, fire-resistant hydraulic fluids, rubber and textile processing aids, and gas compressor lubricants. PAG lubricants are unsuitable for air and refrigeration compressors. There are some gear lubricant and grease applications.

TABLE 2-3 Polyalkylene Glycol Fluids

Viscosity, cSt				Pour point			Flash—C.O.C.		Auto-ignition	
212°F (100°C)	100°F (38°C)	0°F (−18°C)	Viscosity index	°F	°C	Sp. gr. 20/20°C	°F	°C	°F	°C
2.75	11.7	270	83	−70	−57	0.960	325	163	410	210
6.80	35.3	1,400	169	−50	−46	0.983	460	238	572	300
18.5	112	7,100	196	−30	−34	0.997	505	263	671	335
53	365	31,000	219	−10	−23	1.002	515	268	752	400
165	1,100	71,000	281	−20	−29	1.063	545	285	779	415
255	1,970	——	282	40	4.4	1.094	490	254	797	425
2600	19,400	——	414	40	4.4	1.097	620	327	833	445

Phosphate Esters

Phosphate esters derive from the reaction of phosphorus oxychloride with cresylic acids, synthetic alkyl phenols, or certain alcohols. FMC-Houghton, Monsanto, and Stauffer are the largest suppliers, with IMC (Sobin/Montrose) expanding its position.

Fire resistance is unquestionably the most notable property of inorganic phosphate esters. Where combustibility is a hazard, OSHA, Factory Mutual, and other standards increasingly require phosphate ester lubricants. Low volatility and chemical stability are other advantageous properties, but their incompatibility with mineral oil systems is often a disadvantage. Phosphate esters craze and soften certain plastics, coatings, neoprene and nitrile elastomers, and pipe-joint compounds.

Hydraulic fluids and compressor lubricants are the principal uses. Some greases also are used where safety from fire or from very high temperatures is important.

Silicones

Silicones are the reaction products of silicon (from sand or quartz) and different halocarbons, such as alkyl or aryl chlorides. Fluorosilicones are also available. Dow Corning is the major supplier. GE, Union Carbide, and Stauffer's SWS affiliate are becoming more active.

Chemical inertness is the major advantage of silicones. Low flammability and self-extinguishing properties make silicones desirable for many uses. Widespread applicability is limited by their incompatibility with mineral oils and certain additives.

Many specialty applications exist for silicones: moisture-proof seals and lubricants for ignition and electronic equipment and greases for valves and swivel joints exposed to chlorine gas and strong oxidizing or corrosive chemicals. Silicones are used as hydraulic fluids and compressor lubricants. Silicone brake fluids are being used in new systems designed to benefit from their unique properties.

Other Synlubes

Alkylated Aromatics

Alkylated aromatics (dialkylbenzenes, or DAB) were originally co-products of the manufacture of linear alkylbenzene (LAB) "soft" detergent alkylate. Conoco is the major manufacturer. DAB is compounded into low-temperature engine oils, gear lubricants, hydraulic fluid, and grease. Chevron recovers some bottoms from their dodecylbenzene (DDB) "hard" detergent alkylate plant. This highly branched-chain DAB is the base for a high-quality refrigerator-compressor oil and a dielectric fluid.

Cycloaliphatics

Cycloaliphatics are obtained by dimerizing and then hydrogenating the by-product α-methyl styrene from synthetic phenol plants. Monsanto is the only manufacturer today, but Sun also has the technology.

The unique property of these synthesized hydrocarbons is their high traction coefficient. Lubricants compounded from such fluids become shear-resistant semisolids at the high momentary contact pressure in adjustable-speed traction drives. This enables power to be transmitted from one smooth-rolling element to another—without gears, chains, belts, etc. The driving member drives the driven member with essentially no slipping (see Fig. 2-3).

Applications for traction drives include wire-drawing machines, injection molders, filament-payout stands, boring machines, press-roll drives, spring coilers, and gun mills.

Polybutenes

Polybutenes are oligomers of C_4 olefins, principally isobutene. Amoco, Chevron, and Cosden sell some of what they produce, but Exxon and Lubrizol use most of their output as intermediates for lubricant additives.

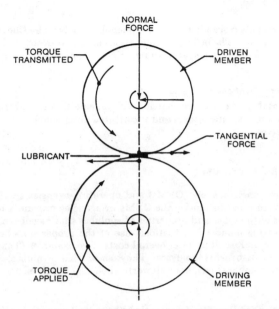

Figure 2-3 Traction coefficient = tangential force/normal force.

Polybutenes combine lubricity with complete burn-off at relatively mild elevated temperatures [600°F (315°C)]. Thus no deposits or stains are left on the hot surfaces. Because of these properties, polybutenes are used in rolling and wire drawing of nonferrous metals. They are also used as lubricants for compressors of nonoxidizing gases, including ethylene and propylene compressors in polyethylene and polypropylene plants.

Halogenated Hydrocarbons

Halogenated hydrocarbons are primarily classified as (1) chlorofluorocarbons, made by Halocarbon, and (2) perfluorinated polyethers, made by duPont and Montedison, and purified further by Bray (distilled fractions).

The extremely inert chlorofluorocarbons are essentially fireproof. They are used in liquid, grease, and plastic form in oxygen plants and in vacuum pumps in gaseous diffusion plants separating isotopes of uranium hexafluoride, an extremely reactive compound. Also they lubricate diaphragm compressors of radioactive gases in nuclear power plants and serve as special fire-resistant hydraulic fluids and suppressants. Because of their high cost, systems to recover the fluid from the gases are generally installed.

Perfluorinated polyethers combine nonflammability with good lubricity. They perform well at temperature extremes and in contact with gases such as oxygen, ozone, fluorine, BF_3, UF_6, etc. Other industrial lubricant applications include furnaces, ovens, chains, rollers, conveyors, and plastic film orienters.

Polyphenyl Ethers

Polyphenyl ethers are best characterized as bis(pheno-oxyphenoxy) benzene, made by Monsanto. This compound possesses extraordinary resistance to degradation by heat, oxygen, radiation, hydrolysis, and chemical attack. It is particularly effective in lubricating electric contacts, especially noble metals. The oil also is used as a very-high-temperature bearing lubricant, in critical heat-transfer systems and in cases where radiation resistance is important.

Silicate Esters

Silicate esters are made from silica sand and phenols or polyols by Chevron, Monsanto, Olin, and Union Carbide. The principal applications are in fire-resistant hydraulic fluids, dielectric coolants, and heat-transfer mediums. Excess moisture causes hydrolysis of silicate esters.

New Developmental Synthetic Functional Fluids

New developmental synthetic functional fluids include perfluorinated triazine and other s-triazines, perfluoro alkylated ethers and substituted polyphenyl ethers, as well as heterocyclic derivatives.

SELECTING THE RIGHT SYNLUBE

Some original equipment makers (OEMs) and operating agencies specify certain synthetic lubricants. In other instances, the OEMs or agencies provide a list of approved synlubes or a list of the types and properties of synlubes that have been found satisfactory. In an increasing number of situations, use of the proper synlube will markedly improve operations and result in lower overall costs. Two examples (Tables 2-4 and 2-5) illustrate simple cost-benefits analyses. They show that synlubes, when properly employed in certain applications, are well worth their higher prices.

TABLE 2-4   Cost Benefits of Synthetic Lubricant: Reciprocating Sodium Hydroxide Metering Pump, Teflon-Impregnated Asbestos Packing

Per year	Packing only (no lubrication)	Packing with fluorosilicone grease
Packing changes	10	1
Labor costs (for changes)	$500*	$ 50*
Packing costs	50	5
Grease cost	——	90
Total maintenance costs	$550*	$145*

Benefits: ● $405* annual savings   ● Less production downtime (worth $$$)

*In 1980 dollars.

TABLE 2-5   Annual Costs (in 1980 Dollars) for 20 Reactor Drives

	Mineral oil	PAO synlube
Lube price, $ -gal	2	10
Lube, changes	Monthly	4 Months*
Requirements, gal	600	200
Cost,	1,200	2,000
Labor, 1 h/change/drive at $30/h;	7,200	2,400
Average repair costs		*
Labor at 16 h/repair,	2,880	960
Materials at $100/repair,	1,200	400
Lost productivity	?	——
Reduced equipment service	?	——
Total tangible costs,	12,480	5,760
Savings,	——	6,720

*Lube change and failure rate is ⅓ of that with mineral oil.

The high costs of new equipment, parts, and labor necessitate keeping machines running better and longer. Such increased efficiency and extended life can be obtained by the judicious use of synthetic lubricants. As costs soar and downtime becomes more expensive, the economics will swing more and more in favor of longer-life, maintenance-saving synlubes—even though their price may be higher than mineral oil's. Where safety is the overriding consideration, cost of the safer synlube will of course remain secondary.

There is no perfect synlube for all applications. Tradeoffs always exist. For some uses mixed-base synlubes are best. Each synthetic has its own individual characteristics. The most appropriate synlube must be carefully selected on the basis of characteristics most suited for the specific conditions and equipment. Endorsements and recommendations of equipment builders and of synlube suppliers should be considered seriously. Only proven products from reliable suppliers should be used. It is wise to review experiences of others in similar situations before trying a new synlube. Synlube papers published by ASLE, ASME, NLGI, and other technical associations usually contain worthwhile information. Brochures and bulletins of synlube suppliers, of course, present much detailed data. More and more independent lubricant suppliers are expanding their lines of synlubes, and are increasingly able to provide high-performance products supported by competent technical service.

Over the longer term, as petroleum supplies continue to decline, more and more lubricants will be synthetic. Starting materials for synlubes may increasingly come from shale oil and coal; also from vegetable and animal oils and other renewable sources, such as biomass. Someday, in the not too distant future, civilization may well run primarily on synthetic, long-lasting, high-performance lubricants.

BIBLIOGRAPHY

Campen, M.: "PAO-based Lubricants in Power Transmission Applications," *6th Annual Conference on Power Transmission,* Chicago, November 1979.

Green, R. L., and Langenfeld, F. L.: "Lubricants for Traction Drives," *Machine Design,* May 2, 1974.

Gunderson, R. W., and A. W. Hart: *Synthetic Lubricants,* Reinhold, New York, 1962.

Manley, L. W.: "New Developments in Synthetic Lubricants," *World Petroleum Congress,* Bucharest, September 1979.

Mueller, E. R.: "Polyalkylene Glycol Fluids and Lubricants," *ASLE Seminar on Synthetic Lubes,* Pittsburgh, February 10, 1977.

O'Connor, J. J., and Boyd, J.: *Standard Handbook of Lubrication Engineering,* McGraw-Hill, New York, 1968, chap. 11, "Synthetic Liquid Lubricants."

Reid, H. F.: *A New Report on Ester Lubricants,* Hatco Chemical Co., Fords, N.J., June 1977.

Smith, R. E.: "Silicone Lubricants for the Chemical Processing Industry," *30th Annual Meeting ASLE,* Atlanta, May 5–8, 1975.

Wolfe, G. F., Cohen, M., and Dimitroff, V. T.: "Ten Years Experience with Fire Resistant Fluids in Steam Turbine Electrohydraulic Controls," *24th Annual ASLE Meeting,* Philadelphia, May 5–9, 1969.

Suppliers' technical bulletins and brochures.

SOURCES OF ADDITIONAL INFORMATION

BASF Wyandotte Corporation, Parsippany, N.J.
Bray Oil Company, Irvine, Calif.
Chevron USA, San Francisco, Calif.
Conoco Oil Company, Houston, Tex.
Dow Corning Corporation, Midland, Mich.
Emery Industries Incorporated, Cincinnati, Ohio.
Ethyl/Edwin Cooper, St. Louis, Mo.
Exxon USA, Houston, Tex.
FMC Corporation, Chicago, Ill.

General Electric Company, Waterford, N.Y.
Gulf Oil Company, Houston, Tex.
Gulf Oil Chemicals Company, Houston, Tex.
Hatco Chemical Corporation, Fords, N.J.
Hercules, Incorporated, Wilmington, Del.
Humko Sheffield Chemical Company, Memphis, Tenn.
Mobil Corporation,
Monsanto Company, St. Louis, Mo.
Olin Corporation, Stamford, Conn.
Stauffer Chemical, Westport, Conn.
Tenneco Chemicals, Incorporated, Piscataway, N.J.
Union Carbide Corporation, New York, N.Y.
Uniroyal Chemical, Naugatuck, Conn.
Witco, Fairfax, Va.

chapter 2-3

Solid Lubricants

by
Rodger C. Dishington
Senior Engineer

Walter D. Janssens
Director, Solid Research

Jack R. Waite
Manager, Engineering Field Services
Imperial Oil and Grease Co.
Los Angeles, California

INTRODUCTION

Solid lubricants are selected solid materials with friction-reducing, low-shear properties. Typical of these materials are graphite, MoS_2 (molybdenum disulfide), PTFE (polytetrafluoroethylene), and mica. They may be used as powders in their natural form, or dispersed in fluids (oils, water) and greases; or they may be added to binders, as pigments to paint, and used as *dry-film* bonded lubricants.

USES

There are many exotic uses for solid lubricants in aerospace, electronics, and instrumentation industries, and machine designers should consider their use wherever conventional lubrication is impracticable. However, solid lubricants play an important role in the practice of industrial "preventive maintenance" as an antiwear, load-bearing component in plant lubricants. Their major benefits are: longer service life of machinery, extended lubrication intervals, and the ability to function in environments hostile to conventional lubricants. They are also used extensively by equipment manufacturers in "lubricated-for-life" components; they are the basis for the 30,000-mile chassis lubrication interval in automobiles.

All machined surfaces, even the ground and polished ones, can be defined in terms of surface roughness, or the dimensions of microscopic high spots called *asperities*. Between the rubbing surfaces of bearings, the lowest coefficient of friction is achieved by maintaining separation of these surfaces by a full *fluid film* (hydrodynamic lubrication). As the oil-film thickness decreases with higher loads, shock loading, or diminishing surface speeds, asperities on opposing surfaces come into contact in *mixed-film* lubrication with a resultant increase in friction.

With increasing severity, a much higher coefficient of friction is experienced during *boundary* lubrication when asperity contact is sufficient to cause microscopic welding and shearing between the two rubbing surfaces. Under the pressures and heat of contact, the solid lubricants may or may not react with the metal substrates, but they do interpose themselves as a barrier coating in which shearing may take place more easily and with less friction than if the bearing metals were shearing in pure contact. As the wear phenomenon is related to the shearing of contacting asperities, wear also may be reduced by the interposing of a solid lubricant barrier.

FORMS AND APPLICATIONS

The useful forms in which solid lubricants are prepared include powders, pastes (compounds), bonded coatings, greases, and dispersions. Powders, burnished into rubbing metal surfaces, have a limited wear endurance and offer virtually no protection from the atmosphere. Pastes are heavy concentrates (up to 65 percent) of solid lubricants in a fluid or grease base; they offer longer, but limited, service life and may include corrosion protection. Bonded coatings may be applied by spray or brush (like paint), by plasma techniques, or by an impingement process if tolerances are critical.

Greases used in antifriction bearings usually contain less than 3 to 5 percent solids, and their particle size is closely monitored, especially for precision and high-speed bearings. Heavy-duty and special-purpose greases may contain up to 25 percent solids. Finished lubricating oils may contain dispersed solid lubricants for improved load-carrying and antiwear characteristics; concentrated dispersions in oil are available as lubricating-oil additives.

The carrier fluids in pastes, greases, and dispersions are not always petroleum-base, but may be water-base or any number of synthetic fluids including glycols, esters, synthetic hydrocarbons, or silicones. Most solid lubricants can tolerate hazardous environments better than most carrier fluids can; thus the carrier is usually selected on the basis of the expected environment. For example: some solids, primarily graphite and MoS_2, are excellent oven-bearing and high-temperature chain lubricants in their powdered form; certain petroleum, glycol, and ester fluids, selected for their favorable volatility are used as media to carry the solids into the rubbing areas to reduce friction and wear even after the fluid has evaporated.

Rubbing metal surfaces are most subject to high friction and wear when they are new or resurfaced, and their ultimate condition and service life depend upon whether they succeed or fail in "running in," or seating properly. Early destruction of contacting asperities or their orderly distortion will determine the true load-bearing area. The less destructive the run-in, the lower will be the magnitude of instantaneous loading. Thus among the most common industrial applications are *wear-in, press-fit,* and *threaded connections.* An expansion on industrial uses appears in Table 3-1.

In addition to the aforementioned materials, hundreds of solid lubricants have been described in technical literature. Those described include metallic oxides and sulfides, soft metals, calcium fluoride, zinc pyrophosphate, talc, and vermiculite. The low-shear characteristics of solid lubricants may be the result of crystal structure, interstitial mat-

TABLE 3-1 Some Industrial Uses of Solid Lubricants

Industrial application	Product form
Wear-in—protection against galling and seizure of newly machined surfaces at start-up and early running-in; examples: gears, slides and ways, cams, valve sleeves, splines, bearings,	Preassembly: powder, paste, bonded coating Initial fill: finished oil dispersion or concentrate
Press-fit—to reduce pressure required, prevent galling, seizure, and possible misalignment; at times clearances are negative; examples: antifriction bearings, splines, keyed shafts, sleeves	Preassembly: powder, paste
Threaded connections, fasteners—to reduce torque loss due to friction, galling, and seizure; promotes optimum uniformity in the tension of assembly bolts and facilitates nondestructive disassembly	Preassembly: pastes (thread compounds), bonded coatings
Life-of-part prelubrication—where maintenance lubrication is impossible, or improbable; examples: enclosed mechanisms, hinges, locks, linkages, instruments, appliances	Preassembly: powder, paste, bonded coating
Lubrication of machine in operation—applied by all conventional systems: drop-feed, circulation, reservoir, air mist; grease gun, automatic grease dispensing systems; for heavy-duty installations and/or extending lubrication intervals and machine service life	Ongoing plant lubrication: finished lubricating oils and greases containing dispersions of solid lubricants
Antiwear, load-bearing additive in lubricants—to fortify conventional lubricating oils and greases used in plant lubrication, initial fills for wear-in, and in units experiencing progressive wear rates	Pastes added to greases, and concentrated dispersions added to oils
Additive to metalworking fluids—to reduce friction, lengthen life, and reduce metal pickup on punching and forming dies and all types of cutting tools	Concentrated dispersions added to metalworking fluids and pastes added to compounds
Reduce fretting (friction oxidation)—to protect against fretting corrosion of metal surfaces under static loads (vibrating) as on bearings, bearing housings, splines, and various press-fit components	Preassembly: pastes and bonded coatings
Antiwear, load-bearing additive in self-lubricated components—to extend life of parts and reduce friction and distortion when blended into rubber, plastics, elastomers, and sintered metals	Powder is added as a component in the raw material for the fabrication of bushings, O rings, etc.
Dry-film lubrication—for use in dusty atmospheres to minimize the adherence of abrasive particles to metal surfaces of open gears, bearings, cams, and slides	Bonded coatings, pastes, greases
High-temperature applications—oven conveyor chains and bearings, kiln-car wheelbearings, mechanical devices operating at temperatures above the capability of fluids	Dispersions in fluids and greases designed to volatilize leaving mainly the solid lubes; some bonded coatings
Equipment exposed to destructive environments—acids, alkali, solvents, detergent, steam, etc. For load-carrying, antiwear characteristics.	Greases, constructed from materials also resistant to the environment

ter, bond strength, or chemical interaction of the surface and the solid. The effectiveness of solid lubricants stems from the almost fail-proof film which they form on moving surfaces, usually beyond the yield strength of the metal asperities. The mechanism of solid lubrication is not dependent on any single property of the lubricant; it is an interdependence of the surface, the solid lubricant composition, the geometry of the particles and the metal surface, and the nature of the processes that occur on or near the bearing surfaces.

CHARACTERISTICS

Solid lubricants composed of two or more materials often combine their most favorable characteristics to provide synergistic lubricating properties that are superior to any single lubricating solid. A brief review of the characteristics of the most commonly used solids can give insight into the lubrication mechanism of solids alone or in combination. The most widely used inorganic and metal base materials are graphite and MoS_2; the most widely used film-forming plastic is PTFE.

Inorganic Lubricants

Graphite and MoS_2 differ in composition, general properties, and type of chemical bonding, but they do have in common their layer-lattice structure. Their characteristic crystal structures are layers of sheets within which the atoms are tightly packed and strongly bonded; but these sheets (laminae) are separated by relatively large distances and held together by weak residual forces. In graphite, a crystalline form of carbon, the distances of atoms within the sheets is 1.4×10^{-8} cm, but between sheets it is as much as 3.4×10^{-8} cm. When under shear, there are strong forces within the graphite sheets, but the forces holding the sheets together are much weaker, allowing them to slip over one another.

A theoretical explanation for the lubricity of molybdenum disulfide is similarly found in its molecular structure. Each lamina of this compound is composed of a layer of molybdenum atoms with a layer of sulfur atoms on each side. The sulfur and molybdenum layers bond tightly but the adjacent laminae interface at their layers of sulfur atoms which form weak bonds between the laminae. The weak sulfur bonds between the laminae form the slippage planes of low shear resistance as between the carbon layers in graphite. Both graphite and MoS_2 display an affinity for metal substrates, and under high loads both have been found to alloy with ferrous metal, forming even greater bonds at the surface.

Under heavy forces (loads) perpendicular to bearing surfaces, these laminae are compressed and oriented parallel to the bearing surfaces and have the strength to resist rupture. The low friction reflects the low resistance of the laminae sliding on one another. Cohesive forces within graphite and MoS_2 are sufficient to allow for self-healing of interruptions in the solid lubricant film.

Plastics

Of the plastic materials used as solid lubricants, PTFE has the greatest affinity for metal surfaces, the lowest internal shear resistance, and the greatest ability to be self-healing. As no plastics have a tight molecular structure or laminar nature, they are unable to take the heavy loads carried by graphite or MoS_2. At loads up to 25,000 lb/in^2 (1760 kg/cm^2) however, PTFE has the lowest coefficient of friction of all solid lubricants.

Common Characteristics

A primary characteristic of these three leading solid lubricants is their ability to lubricate as dry powders. Almost all other solid lubricants require some addition of fluid or grease to lubricate for any period without noise and increasing friction. Nor do other solids have the excellent film-forming and self-healing characteristics of these three which often form the base to which other solids are added. Some of the following are the characteristics which are sought by the use or addition of other solids: particle geometry suited to antiseize or wear-in of coarse surfaces, improved protection against fretting wear, improved bulk-film load-carrying capability, and the thickening of some fluids for greases. Some white solid lubricants, e.g., zinc sulfide, are used alone where a white or colorless lubricant is required for appearance. Solids, even though not laminar, still become interposed between rubbing surfaces when dispersed in fluids or greases.

It is again the superior properties of graphite, MoS_2, and PTFE that allow for their use in very adverse environments. In high-temperature operations graphite is commonly used to 800°F (425°C) and intermittently to 1200°F (650°C). In vacuum at high temperature, or in high vacuum, graphite loses a water-vapor layer which is found naturally adsorbed at the surfaces of the laminae and is felt to be largely responsible for the weak attraction or bond between them. Loss of the water vapor results in a significant increase in friction and a tendency toward abrasion. On the other hand, MoS_2 has a lower coefficient of friction in the absence of water vapor and it can withstand very high vacuum, but it is not as resistant to oxidation as is graphite. MoS_2 performs well up to 650°F (345°C). Above this temperature, in air, its oxidation rate increases until at 750°F (400°C) it oxidizes quite rapidly. PTFE will perform well in vacuum and at temperatures up to 600°F (315°C) at which it begins to decompose.

Solvents and chemicals provide some of the environments most destructive to conventional lubricants. Again, the leaders are the most impervious of all solid lubricants, being virtually unaffected by direct contact. Graphite and PTFE are nearly indestructable by any chemicals, and MoS_2 is relatively unaffected except by hot strong oxidizing acids like aqua regia and chromic acid. At times, the removal of MoS_2 from metal surfaces requires strong caustic cleaners and abrasive action. Special solvent-, chemical-, and detergent-resistant greases containing solid lubricants require also the careful selection of fluid carriers and thickeners.

SUMMARY

There is a need for a solid lubricant wherever a hydrodynamic oil film cannot be sustained. Whether the film becomes insufficient because of high pressure or shock loading, elevated temperatures, or exposure to solvents and chemicals or is depleted by the extension of lubrication cycles, solid lubricants can extend service life by inhibiting asperity welding (wear), galling, and the ultimate seizure of bearing metals. All bearings are in boundary lubricating conditions during stop-start operations since full fluid-film lubrication requires motion to propagate.

BIBLIOGRAPHY

Acheson, E. G.: U.S. Patent 813,426, February 5, 1907.
ASLE Proceedings, International Conference on Solid Lubrication, Chicago, 1971.
ASLE Proceedings, 2nd International Conference on Solid Lubrication, Chicago, 1978.
Bowden, F. P., and D. Tabor: *The Friction and Lubrication of Solids,* Oxford, New York, 1950.
Braithwaite, E. R.: *Solid Lubricants and Surfaces,* Pergamon, New York, 1964.
Braithwaite, E. R.: *Lubrication and Lubricants,* Elsevier, New York, 1967.
Campbell, M. E.: *Solid Lubricants, A Survey,* NASA SP-5059 (01), 1972.
Deoine, N. J., E. R. Tourson, J. P. Cerini, and R. J. McCartney: "Solids and Solid Lubrication," *Lubrication Engineering,* January 1965.
E. Hall: U.S. Patent 2,700,623, April 26, 1950, and U.S. Patent 2,703,768, April 21, 1954.
McCabe, J. T.: "Molybdenum Disulfide—Its Role in Lubrication," reprint of a paper presented at International Industrial Lubrication Exhibition, New Hall, Westminster, London, March 8–11, 1965.

Neale, M. J. (ed.): *Tribology Handbook,* Wiley, New York, 1973.
Notes on Solid Lubricants, Seminar Proceedings at Rensselaur Polytechnic Institute, 1966.
Solid Lubricants, NTIS/PS 75/715, 78/0816, 78/0817, 78/0818, 78/0819, NTIS Bulletins, U.S. Dept. of Commerce, Washington, D.C.
Stock, A. J.: "Graphite, Molybdenum Disulfide and PTFE—A Comparison" *Lubrication Engineering,* August 1963.
Ubbelohde, Q. R., and F. A. Lewis: *Graphite and its Crystal Compounds,* Oxford/Clarendon, 1960.
Waite, J. R., and W. D. Janssens: "Use of Inorganic Solids in Lubricating Oils," Presented at ASLE Annual Meeting, Cleveland, Ohio, May 6–9, 1968.
Youse, E. L., NLGI Spokesperson: *Characteristics and Selection of Graphite as a Lubricant,* January 1962, Kansas City, Mo.

chapter 2-4

Lubrication Systems

by
Anne Bernhardt
Staff Engineer
Gulf Research and Development Company
Petroleum Products Department
Pittsburgh, Pennsylvania

INTRODUCTION

An effective lubrication system is any system or device which dispenses the correct lubricant, at the correct point, in the correct amount, at the correct time. Systems may vary from hand oiling to a complicated centralized system. Escalating costs and the development of high-speed precision machinery are necessitating changes in plant lubrication practices.

ALL-LOST SYSTEMS

All-lost or once-through lubrication (Fig. 4-1) systems are those in which the lubricant is used only once. Manual all-lost systems such as hand oiling, individual grease fittings, wick lubricators, oil cups, and drop-feed oilers are rapidly becoming a part of history. These systems are inexpensive to install but require close attention on the part of the operator to ensure that each point is relubricated on a regular basis for adequate lubrication.

The most common all-lost systems in use today are automatic. Mist systems and mechanical force-feed lubricators are common examples. The reason for their popularity

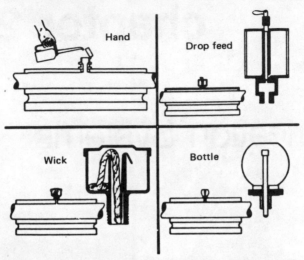

Figure 4-1 Once-through oiling *(U.S. Steel Corporation, reprinted with permission from "The Lubrication Engineers Manual," Copyright 1971).*

is their ability to lubricate more than one point on a machine from a central reservoir and to automatically dispense the lubricant in metered amounts at the point of application.

Mist systems rely on compressed air to atomize the oil into fine droplets and deliver it through pipes to the point of application. Frequently used to lubricate bearings and gears, the air passing over the part assists the oil in carrying off heat and preventing the entry of dirt.

Mechanical force-feed lubricators were originally designed to deliver oil to cylinders but are no longer limited to that application. They consist of small pumping units mounted on a common shaft which take oil from a reservoir through pipes to the point of application. The driving mechanism may be an electric motor or some moving part of the machine being lubricated. Each pumping unit may be set to deliver the precise amount of oil which is required at a particular application point. Both the mist system and the mechanical force-feed lubricator require little maintenance beyond ensuring that no lines are plugged.

OIL-RESERVOIR SYSTEMS

Unlike all-lost systems, oil-reservoir or self-contained systems reuse the same oil over and over again. These methods depend on a common housing containing the oil and the parts to be lubricated (Fig. 4-2).

Gears and cylinders lubricated by these methods usually depend on the splashing action of one or more moving parts dipping into the pool of oil at the bottom of the housing.

Bearings lubricated by self-contained systems may be splash-lubricated or rely on a ring, chain, or collar to dip into the oil and carry the oil to the top of the journal. Collar bearings are used at higher speeds than rings or chains since the latter will tend to slip excessively at high speeds, precluding adequate lubrication.

To ensure adequate lubrication it is important that the oil be maintained at the proper level. Insufficient oil could result in a lack of lubrication, while overfilling can cause foaming and temperature buildups due to excessive churning.

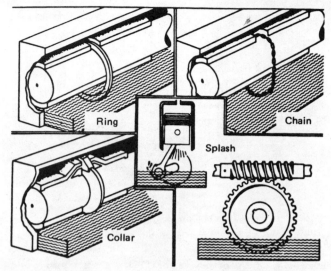

Figure 4-2 Oil reservoirs *(U.S. Steel Corporation, Reprinted with permission from "The Lubrication Engineers Manual," Copyright 1971).*

CENTRALIZED SYSTEMS

Like oil-reservoir systems, centralized systems use the oil over and over again. They can range from a simple reservoir, pump, and return-line setup to complex systems with electronic controls, servo valves, heat exchangers, and filters.

Depending on the complexity of the system, costs vary greatly. The cost-effectiveness of centralized systems depends heavily on the length of time the fluid can remain in circulation before it needs to be changed.

To ensure maximum fluid life, oil-reservoir temperatures should be controlled as well as the amount of contaminants in the oil. If petroleum oils are used, reservoir temperatures should be maintained between 110 and 130°F (43 and 54°C) for optimum fluid life. Synthetics can be operated at somewhat higher temperatures. Reservoir temperatures can be controlled through the use of heat exchangers and proper reservoir design.

The oil reservoir should be large enough to allow the oil to rest for a minimum of 15 min before being recirculated. It should be baffled to ensure that returning oil is not immediately pumped back into circulation. This rest time in the reservoir allows the oil to drop out contaminants and dissipate the heat it has picked up while in circulation.

The oil-reservoir's fluid level is also very important to the trouble-free operation of circulating systems. If the suction line is not completely submerged in oil at all times, pump cavitation could result. Also, it is important that the return line be submerged in the oil to reduce air entrainment and thus prevent foaming problems which could occur if the returning oil is allowed to splash into the reservoir. Figure 4-3 is an example of a properly designed reservoir.

Monitoring and Warning Devices

A properly designed centralized or automatic lubrication system can provide effective equipment lubrication with very little human intervention. The main drawback is the risk of catastrophic equipment failure which may result when a malfunction occurs. To prevent this, monitoring and warning devices are installed to alert operators to lubrication malfunctions. These may be bells, sirens, flashing lights, automatic equipment shut-

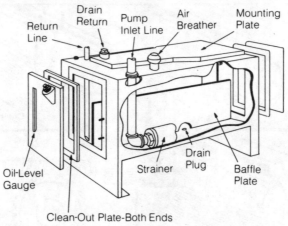

Figure 4-3 Typical circulating system reservoir.

down, or telltale indicators. All of these can be effective if maintained in proper operating condition.

CRITERIA FOR SELECTING A LUBRICATION SYSTEM OR DEVICE

A lubrication system should be selected with one purpose in mind: *to get the proper quantity of the correct lubricant where and when it is needed.* Before purchasing or installing a new lubrication system it is recommended that the application be studied to aid in the selection of the most economical and effective system for a particular application. Factors to be considered include:

Equipment Considerations
- The components to be lubricated
- The lubricant to be applied
- The lubrication-point accessibility
- The number of lubrication points the system is expected to service

Operation Condition Considerations
- Equipment speeds
- Operating temperatures
- Expected relubrication intervals

Economic and Plant Practice Considerations
- Plant's past experience with various types of lubrication systems
- Available capital
- Staff available to maintain and monitor systems
- Downtime costs of the equipment

Oil companies, equipment builders, and lubricant-dispensing-equipment suppliers are available to assist in the design and selection of lubrication systems. A list of some of the better-known manufacturers of lubricant-dispensing equipment follows.

LUBRICANT-DISPENSING EQUIPMENT MANUFACTURERS

Alemite Division, Stewart-Warner Corp., Chicago, Ill.

Bijur Lubricating Corp., Rochelle Park, N.J.

Farval Division, Fluid Control Division, Eaton Corp., Cleveland, Ohio

Lincoln St. Louis, A Division of McNeil Corp., St. Louis, Mo.

Lubriquip, Cleveland, Ohio

Madison-Kipp Corp., Madison, Wis.

C. A. Norgren Co., Littleton, Colo.

Oil-Rite Corp., Manitowoc, Wis.

BIBLIOGRAPHY

Bailey, Charles A., and Joseph S. Aarons (eds.): *The Lubrication Engineers Manual,* United States Steel Corp., Pittsburgh, 1971.
Brewer, Allen F.: *Basic Lubrication Practice,* Reinhold, New York, 1955.
Brewer, Allen F.: *Effective Lubrication,* Robert E. Krieger, Huntington, N.Y., 1974.

section 3

Corrosion and Deterioration of Materials

chapter 3-1

Causes and Control of Deterioration

by
Philip H. Maslow, P.E., FCSI
Consultant
Chemical Materials for Construction
Brooklyn, NY

INTRODUCTION

All materials deteriorate. External building materials and assemblies are no exception to this law. Scientific observation shows that some materials deteriorate at a faster rate than others, depending on a set of controlling conditions, or causal factors. Often, these conditions can be modified to alter the rate of deterioration to a preferred or acceptable level.

The *total performance* of a building refers to the ability of both external and internal materials to fulfill their intended function over the useful life of the building. The performance of the external materials has special significance here in terms of maintenance costs and replacement costs. The differences between external and internal performance are tied into the differences between the effect of wear caused by use and that of wear by the environment. Experience with new building types has shown the need for considerable caution because one of the problems in the use of new materials is their breakdown or failure, particularly on the exterior.

Deterioration is almost the antonym of durability. Durability is essentially a *state* whereas deterioration is a *rate*. Deterioration is the result of corrosion, chemical action, atmospheric pollution, structural movement, and user wear and tear. The environment comprises a set of elements such as air-temperature, rainfall rate, wind speed, pollution level, and smoke concentration. These elements have special significance because they are the ones over which the least control can be exercised and have the widest range of effects on exterior building materials.

Standard test results indicate that the type of atmospheric pollutants normally found in urban areas accelerate the breakdown of concrete, brick, mortar, paints, and various metals and plastics. Some of the pollutants most commonly encountered in urban and industrial areas are carbon dioxide, carbon monoxide, sulfur dioxide, sulfur trioxide, and nitrous oxide. These pollutants react with moisture in the atmosphere to form carbonic, sulfurous, sulfuric, and nitric acids, respectivly. And these corrosive acids act upon all building materials.

Because of new zoning regulations and current design trends, many new commercial buildings, and even industrial buildings, have plazas, walks, and landscaped areas containing reflecting pools, fountains, and, in some cases, huge sculptures. These areas can be paved with materials varying from concrete, paving brick, bituminous concrete shapes, and terrazzo to natural minerals such as slate, bluestone, granite, and Travertine marble.

For walls there is a choice of brick, poured-in-place concrete, precast concrete, cast stone, ceramic veneers, marble, limestone, granite, etc. In addition, there are window walls combined with masonry and curtain walls, which are all glass with metal frames.

Plastics have entered the construction field in recent years to such an extent that millions of pounds of plastics are now used annually in a variety of installations. Dozens of test methods for evaluating plastics indicate how extensively they are utilized in construction: these include tests for abrasion resistance, bearing strength, brittleness, crazing resistance, deformation under load, flame resistance, flexural strength, indentation hardness, mildew resistance, shatterproofness, shear strength, tensile strength, thermal expansion, and water absorption.

The history of metals in architectural applications extends over hundreds of years: similarly, large-scale use of metals on building surfaces has a long tradition. Today, the demands of technology have resulted in the development and production of many alloys covering an extensive range of properties. Fundamental to the deterioration of metals is the phenomenon of corrosion. Ferrous metals are particularly subject to corrosion. Copper, lead, and zinc corrode, initially, on exposure to the atmosphere, but the product of their corrosion becomes a protective layer which serves to isolate the metal from further attack. Stainless steel and aluminum become passive upon exposure to the atmosphere. These metals acquire a surface film which isolates the metal and effectively protects it from corrosive attack. To protect aluminum even further, it is subject to modification by alloying with various metals and by anodizing. This is a process in which the thickness of the surface film is increased by electrolytic oxidation. By including various organic dyes in the electrolytic bath, the aluminum may be anodized in a range of colors. Coatings to protect carbon steel and other sensitive metals have also been developed.

Water is one of the prime elements in the composition of the Earth. Although it is basic to the sustenance of life, it is also a highly destructive material and has a great effect on the durability of the exterior of a building. The sources of water problems and the resultant damage to a structure extend from the roof, through the fabric of the building, to the foundation below grade.

The causes of water infiltration at the roof area can be traced to several sources. The parapet is a common source of water infiltration because of the lack of proper flashing details, poor mortar joints, poor coping stones, and poor sealants. Roof expansion joints, if not properly designed or installed, can lead to splitting of the roofing membrane and to subsequent water penetration. The detailing and construction of the roofing membrane, beginning with the vapor barrier, the insulation, the plies, the flood coat, and the gravel ballast, all have a bearing on the durability of the entire roofing system.

The envelope of a building is also subject to penetration by water. Rain does not affect the roof area exclusively; it can wreak havoc on all sides of a building, especially, when there is a wind factor behind it. The sources of infiltration vary with the kind of wall fabric. There are absorption factors to consider, and these factors vary with the construction material. Joints and sealants, flashing and flashing details, and weep systems—all have a bearing on waterproofing and/or water infiltration.

The effects of water damage to the exterior are both unsightly and damaging structurally. Masonry can spall, especially in a freeze-thaw cycle or in a continuous saturation process. Rusting of structural steel members can result in serious deterioration of beams, relieving angles, and lintels. Of course, the damage from water entering the interior affects plaster, paint, drywall, ceilings, floors, carpets, curtains, and furniture.

Modern curtain walls and window walls are equally vulnerable to water infiltration. Water may come through stack or expansion joints in column covers, mullions, or tracks. Water may enter because of improper metal-to-metal joinery, improper joint design, poor sealant details, poor shop seals and/or field seals, and poor gutter or weep systems.

Glazing systems in every kind of window detail, including skylights, are also subject to water infiltration. One source of infiltration can be a faulty glazing system. Poor workmanship, inadequate materials, and poor design of the glass surrounds are other major reasons for water infiltration.

This chapter explains the causes of deterioration of concrete, masonry, metals, wood, and plastic and outlines methods of maintenance, repair, and rehabilitation.

CONCRETE*

CAUSES FOR DETERIORATION OF CONCRETE

The visual symptoms of concrete deterioration are cracking, spalling, and disintegration. Each is obvious and may occur individually or in any combination. Most of the basic causes for the deterioration of concrete are noted in the following text.

Inadequate Design Details

Deterioration is often caused by design deficiencies such as nonworking joints between precast members; poorly sealed joints; inadequate drainage at foundations, at horizontal surfaces such as roofs and tops of walls and sills, and at inadequately placed or incomplete weep holes; unanticipated shear stresses in piers, columns, or abutments; incompatibility of materials (such as concrete in contact with aluminum or concrete containing calcium chloride in contact with reinforcing steel); neglect of the cold flow factor in the deformation of concrete under stress; and inadequate provision for expansion joints, control joints, and contraction joints.

Construction Operations

Problems during construction include settlement of the subgrade, movement of formwork that is inadequately designed or built, vibration of concrete during set, segregation of the concrete with heavy aggregates settling, and removal of forms before concrete has achieved minimal strength. Segregation can be minimized by proper water-cement ratios, which will produce concrete of proper consistency that is "placed" rather than "poured." Proper compaction by vibration is mandatory.

Drying Shrinkage

Contraction caused by loss of moisture from green concrete may be minimized by using proper water-cement ratios, proper vibration, adequate reinforcement, and effective curing membranes or other curing methods.

Temperature Stress

Variations in atmospheric and internal temperatures can cause enough stress in the concrete to cause cracking. Daily and seasonal variations in atmospheric temperature and the coefficient of thermal expansion of concrete must be considered when designing and locating joints.

Moisture Absorption

Premature spalling can be caused by moisture absorption. De-icing salts accelerate this phenomenon. The use of proper sealers and air-entrained concrete minimize this type of disintegration.

Corrosion of Reinforcing Steel

Corrosion of reinforcement may be attributed to the types of cements, sands, and aggregates used in the concrete mix, as well as to the type and amount of water. Under some conditions, admixtures, such as accelerators and water-reducing agents, can contribute to corrosion of reinforcing steel.

*This material adapted with permission from *Plant Engineering*, October 13,1977, pp. 215–222.

Chemical Reaction

Many types of chemicals are deleterious to concrete. Salts in soil, as well as bacteria, can affect concrete. Saltwater is highly corrosive to steel in concrete. Carbon dioxide and other products developed by the exhaust systems of internal-combustion engines add acids to the atmosphere that attack concrete.

Wear

Erosion or abrasion are common factors that affect concrete, particularly in traffic areas. Hydraulic structures, such as dams, are subject to cavitation erosion because of rapidly flowing water.

Impact

Floors in industrial plants are particularly subject to damage by heavy impact. Heavy reinforcement, combined with a concrete of high compressive strength are the usual methods for providing impact resistance.

Weathering

All concrete exposed to the atmosphere is subject to attack from the elements, from ultraviolet light, and from the myriad chemicals found in the atmosphere and in industrial plants.

DIAGNOSING DURABILITY PROBLEMS

Deterioration may be the result of corrosion, chemical action, atmospheric pollution, structural movement, or wear and tear. Diagnosing the causes of deterioration in a concrete structure may be a matter of eliminating possibilities. The process for identifying possible causes should include:

- Checking for errors in basic design. The knowledge and experience of an architect or a structural engineer will usually be required.
- Relating basic symptoms (cracking, spalling, and disintegration) to possible causes.
- Checking first for readily identifiable causes, such as corrosion of the reinforcing steel where the concrete has broken away to reveal the rusty bars and concrete that has been damaged by impact or shock, revealing reinforcing steel that is not yet corroded.
- Making a specific investigation of the concrete mix design, methods of construction, curing methods, time of year of installation, reputation of the contractor, etc.
- Analyzing clues. Each of the principal symptoms should be examined individually. If the concrete is disintegrating, check for chemical attack, erosion, or general weathering. If the concrete is spalling, check for variations in internal temperature, chemical and electrolytic corrosion of the reinforcing steel, and poor design details as well. If the concrete is cracking, check all the foregoing factors and for variations in atmospheric temperature and accidents during construction.
- Using the process of elimination. If initial construction or design details are responsible, redesign and corrective measures may be necessary. If external causes are responsible, protective measures must be exercised. Protection and maintenance are necessary to maintain the structure in a sound condition.

There are several accepted procedures and methods for repairing cracks in concrete structures, spalled concrete floors and pavements, and disintegrated concrete members. Table 1-1 outlines several types of materials for such repairs. The materials described can also be used to repair masonry members.

TABLE 1-1 Latex and Epoxy Adhesives and Bonding Agents for Concrete

	Latices			Epoxies				
	Acrylic	Polyvinyl-acetate (nonre-emulsifiable)	Butadiene-styrene	Epoxy-polysulfide binder only	Binder with sand	Epoxy-polyamide Binder only	Epoxy-polyamide Binder with sand	Epoxy–coal tar binder only
Appearance	Milky white	Milky white	Milky white	Light straw to amber	Light straw to amber	Light straw to amber	Light straw to amber	Black
Solids content, %	45	55	48	100	100	95 to 100	95 to 100	100
Reference specifications	MIL-B-19235	MIL-B-19235	MIL-B-19235	MMM B-350A; AASHTO M-200	MMM G-650A; AASHTO M-200	AASHTO M-200	AASHTO M-200	AASHTO M-200
Chemical resistance Acids Alkalis Salts Solvents	Fair Very good Very good Fair to good	Fair Very good Very good Fair to good	Fair Very good Very good Fair to good	Excellent Excellent Excellent Excellent	Excellent Excellent Excellent Excellent	Excellent Excellent Excellent Excellent	Excellent Excellent Excellent Excellent	Excellent Excellent Excellent Excellent
Compressive strength, lb/in² (2-in cubes; ASTM C 109)	3200 to 4100	3400 to 3600	3300 to 4000	8000 to 10,000	12,000 to 15,000	6000 to 8000	10,000 to 13,000	3000 to 4000
Tensile strength, lb/in² (1-in briquettes; ASTM C 190)	580 to 615	340 to 450	450 to 580	—	—	—	—	—
Tensile strength, lb/in² (ASTM D 638)	—	—	—	3000 to 3500	—	—	3500 to 4000	400 to 800
Tensile elongation, % (ASTM D 368)	—	—	—	2.5 to 15	—	6 to 25	—	35 to 40

Property								
Flexural strength, lb/in² (bar; ASTM C 348)	950 to 1400	1000 to 1250	1250 to 1650	—	—	—	—	—
Compressive double-shear strength, lb/in² (MMM G-650A)	—	—	—	900 to 1000	700 to 1000	400 to 500	500 to 650	300 to 400
Application notes	Suitable for indoor and outdoor exposure on concrete, steel, wood, thin section toppings; shotcrete, plaster bond within 45 to 60 min; not suitable for extreme chemical exposure Do not use with air entrainers	Not for conditions of high hydrostatic head Can be used with accelerators, retarders, and water-reducing agents, but not with air entrainers	Not for constant water immersion Not for use with air entrainers or accelerators	Suitable for filling cracks in concrete to bond both sides of crack into an integral member; preparation of epoxy mortars by adding sand For maximum chemical and physical properties; highest cost; not for use on surfaces treated with rubber or resin curing membranes, dirty surfaces, weak concrete, or bituminous surfaces	Suitable for bonding hardened concrete and other materials to hardened concrete; setting dowels; bonding plastic concrete to hardened concrete; bonding skid-resistant materials to hardened concrete For maximum chemical and physical properties; highest cost; not for use on surfaces treated with rubber or resin curing membranes, dirty surfaces, weak concrete, or bituminous surfaces	Suitable for filling cracks in concrete to bond both sides of crack into an integral member; preparation of epoxy mortars by adding sand For maximum chemical and physical properties; not for use on surfaces treated with rubber or resin curing membranes, dirty surfaces, weak concrete, or bituminous surfaces	Suitable for bonding hardened concrete and other materials to hardened concrete; setting dowels; bonding plastic concrete to hardened concrete; bonding skid-resistant materials to hardened concrete For maximum chemical and physical properties; not for use on surfaces treated with rubber or resin curing membranes, dirty surfaces, weak concrete, or bituminous surfaces	Suitable for preparation of epoxy mortars by adding sand; bonding skid-resistant materials to hardened concrete; membrane between asphalt and concrete For resistance to grease, oil, gasoline, and traffic; use on bituminous concrete; lower cost applications of nonskid membranes; not to be used for bonding new wet concrete to old or where black color will be undesirable

REPAIRING CRACKS

It has often been said that concrete is destined to crack. The purpose of proper design and the objective of the design engineer are to minimize cracking without the expectation of eliminating it. When cracking does develop, it is important to determine the basic causes, as well as its extent, before deciding on the method of repair.

The surface of concrete will often exhibit shrinkage cracks—these are not necessarily defects, but they are aesthetically undesirable. They are usually attributed to the use of a high-slump concrete that contains excessive water. Cracking occurs when excess moisture leaves the concrete too soon—before the concrete has sufficient tensile strength. Such cracks can be minimized by using a sheet membrane or curing compounds. Although this type of cracking may not necessarily lead to problems, and it may often be ignored, applying a sealer based on a synthetic-rubber compound to protect the concrete from further damage is prudent.

Cracks can be described as active or dormant. An active crack will open and close with changes in temperature and with cyclic movement of the structure. Dormant cracks may not go through such movement, but they may still leak, collect dirt, interfere with traffic, etc. A structural crack may usually be attributed to inadequate structural design, insufficient strength (material composition), or poorly designed joints. If joints are not provided in concrete slabs, the concrete will create its own joints by cracking, and these cracks will continue to develop until the concrete member comes to equilibrium.

Active cracks may be sealed with an elastomeric sealant. Dormant cracks may be sealed with a fluid epoxy sealer that can be pumped into the crack or allowed to flow in by gravity. An epoxy sealer of 100 percent solids content will seal the crack without shrinkage and will join the crack faces to re-form a monolithic structure. This seal will be strong enough to resist further cracking. However, should stress still occur, cracking will take place somewhere else in the structure.

Horizontal cracks may be filled by simply pouring in a liquid epoxy sealer until it overflows, indicating that the crack has been filled.

Vertical cracks may be filled with a liquid epoxy sealer. First, the face of the crack is sealed with a fast setting epoxy compound, which is allowed to cure thoroughly. Small holes are then drilled into the crack through the epoxy seal, and nipples are installed in these holes and bonded with the same fast-setting epoxy. The low-viscosity epoxy sealer is then injected into the lowest nipple. Pressure is maintained until liquid begins to seep out of the next higher nipple. These two nipples are then plugged and the same procedure is resumed with the third nipple. This operation is continued until the entire crack is filled. After the sealer has cured, the nipples may be cut off flush with the concrete surface.

The tensile strength of a cracked concrete member may also be restored by stitching. U-shaped iron rods known as stitching dogs are inserted in drilled holes to transfer stress across the crack. Holes are drilled on both sides of a structural crack, far enough away not to cause additional breaks but not in parallel position, which would produce a plane of weakness. The legs of the stitching dogs are designed to be long enough to provide adequate pull strength. After the legs of the stitching dogs are inserted into the holes, the holes may be grouted with a nonshrinking grout. The crack itself should be sealed with an elastomeric or an epoxy sealer to prevent water from entering.

The elastomeric sealants should be based on polysulfide or urethane rubbers (preferably a two-component formulation in traffic grade) and comply with Federal Specification TT-S-00227E. The epoxy sealers may be epoxy-polysulfides (complying with Corps of Engineers Specification MMM B-350A) or epoxy-polyamides (complying with Specification M-200-65 of the American Association of State Highway and Transportation Officials). The epoxy compounds are described in Table 1-1.

In crack repair, certain procedures are detrimental to the successful repair of the crack.

• Filling cracks with new mortar will result in further cracking.

- Placing a topping over a crack to seal it, unless the topping is elastomeric, will inevitably result in the crack passing through the topping itself.
- Repairing a crack without relieving the restraints that caused it will cause cracking elsewhere.
- Repairing a crack with exposed reinforcing that has begun to corrode should not be done until the steel has been cleaned and protected with a rust-inhibitive paint.
- Burying a joint that has been repaired prevents frequent inspection to determine whether further failure has occurred.

REPAIRING SURFACES OF PAVEMENT AND SLABS

A spalled concrete surface is the beginning of continued disintegration of the concrete. If the cost of replacement is at all comparable with the cost of repair, replacement is recommended. Replacement is also recommended when changing the level of the final surface is impractical. Otherwise, the concrete can be resurfaced with new concrete, epoxy topping, latex mortar, or iron topping.

Resurface with New Concrete

When there is no problem in changing the level of the surface, it may be resurfaced with new concrete. It may be laid with or without bonding, although bonding is always recommended. The old surface should be prepared by removing all loose material and contaminants.

A simple bonding can be made by scrubbing a neat cement slurry (a mixture of straight portland cement and water) into the surface. The new concrete may then be placed with the expectation of a good bond.

Concrete placement should follow recommended practice. The concrete mix should have a water-cement ratio of 0.50 or less and a slump of 3 to 4 in. It is also recommended that reinforcement be embedded in the middle of the topping slab, rather than allowed to lie at the bottom of the slab.

Finishing procedures will depend on the required surface: use a wood float finish for a regular surface and steel troweling for a hard, smooth, sealed finish. The slab should be cured by covering it with sheet materials or liquid curing membranes. Concrete overlays of this type should be at least 2½ to 3 in thick.

A more positive bond may be achieved by using an epoxy bonding agent, either a filled epoxy-polysulfide or an epoxy-polyamide bonding agent (see Table 1-1). The epoxy bonding agent may be a proprietary formulation or one meeting a specification of a governmental agency. The bonding agent must be thoroughly mixed and applied by brush, broom, or spray. Do not apply more bonding agent than can be covered with new concrete while the bonding agent is still tacky. If the film of bonding agent has set, apply fresh epoxy adhesive before applying new concrete.

Shearing of the new concrete course from the base slab at the bond line is unlikely if the bonding-agent film is still tacky when the topping is placed. However, should a crack develop in the base slab, there is a definite possibility that this crack will transfer through the epoxy bonding-agent glue line and through the new wearing course.

New urethane bonding agents, which are often used between slab waterproofing membranes, can inhibit crack transfer from the base slab to the topping. The urethane membrane is also a two-component system that is applied in the same way as an epoxy bonding agent. Being elastomeric, the urethane membrane can absorb more of the stresses that are set up as the concrete overlay cures and contracts. It is also flexible enough to stretch and bridge cracks that develop in the substrate, as well as in the wearing course, thereby preventing transfer of cracks.

A relatively inexpensive bonding agent that may be used to ensure a positive bond of a new concrete overlay is a latex-reinforced cement slurry grout. Portland cement and sand (in a ratio of 1:3 by volume) are combined with a gauging liquid that is a mixture of equal volumes of water and an acrylic or polyvinyl acetate latex emulsion, as described in the table. This emulsion must be nonre-emulsifiable. The gauging liquid, based on the latex blend, is added to the cement-sand powder until a creamy paste is developed. This paste is scrubbed into the surface of the base slab, covering the surface thoroughly. New concrete is then placed on this bond line, which is approximately ¹⁄₁₆ in thick.

It is recommended that the latex slurry not be applied too far ahead of the placement of the concrete.

The latex bonding agent should not be used by itself because too much could be absorbed by a porous substrate, leaving a minimal thickness in the glue line. It may also dry too quickly to form an effective bond by the time new concrete is applied. It is best used in the form of a slurry grout.

Resurface with an Epoxy Topping

When the level of the surface must be kept close to the original elevation and a chemically resistant and tougher wearing surface must be provided, an epoxy topping may be the answer. After the base slab has been thoroughly cleaned by sandblasting, grinding, or acid etching, it is rinsed and allowed to dry. The epoxy mortar is supplied as a three-package proprietary system consisting of a base epoxy resin, a catalyst, and a measured quantity of dry, salt-free sand. The ingredients are mixed together to form a mortar which is applied to a thickness ranging from ⅛ to ⅜ in, screeded, and troweled smooth (Fig. 1-1). Most systems also include a primer, based on an epoxy system, to be applied before the epoxy mortar.

Figure 1-1   Surface of concrete slab or pavement can be repaired with thin trowel-applied overlay of sand-filled epoxy resin.

Resurface with a Latex Mortar

Although less resistant to chemicals than epoxies, latex mortar toppings also allow resurfacing without significantly changing surface elevation. The mortar may be made by blending 1 part (by volume) portland cement with 3 parts mason's sand. Latex is then added at the rate of 10 percent solids based on the cement content. Enough water is added to make a trowellable mixture. A typical formulation would be:

Portland cement (Type I)   1 bag (94 lb)
Mason's sand   3 bags (300 lb)

| Latex emulsion (50 percent solids) | 2 gal |
| Water | 4–5 gal |

Before the latex mortar is applied, a latex cement slurry should be applied as a bonding agent. The slurry may have the same latex as that used in the mortar topping. Latex mortar is best spread and leveled with a wood float rather than a steel trowel, because the latex tends to rise to the surface and cause a drag on the steel trowel.

After the latex mortar is finished, a sealer should be applied. The sealer, which functions as a curing membrane and protects the latex mortar against contamination from grease, oil, and deicing salts, may be based on a chlorinated rubber, a butadiene styrene rubber, or a methyl methacrylate resin. This type of sealer may be applied to all new concrete.

Resurface with an Iron Topping

Areas subject to very heavy traffic, particularly to steel-wheeled vehicles, require an extra durable surface such as that produced by an iron topping. It is usually applied in thicknesses of 1 in. Specially graded iron particles are substituted for a major percentage of the sand aggregate in a mortar or concrete mix. This topping may be bonded to a substrate, using a neat cement slurry, a latex cement slurry, or a urethane or epoxy bonding agent. This iron topping is usually available in a ready mixed form or can be blended at the jobsite following the recommendations of the supplier of the iron particles.

REPAIRING DISINTEGRATED CONCRETE MEMBERS

Methods and materials for repairing columns, beams, piers, and precast concrete panels will generally be similar to those used in repairing concrete pavements. However, because such concrete members are either load-bearing or an integral part of a structure, it is not always possible to remove and replace them. Therefore, repairs must be made to the existing structure using the best means available.

Before any repair or replacement of concrete sections is attempted, steel that is exposed by spalled or disintegrated concrete must be treated to prevent further corrosion, which may have been the initial cause of the disintegration. Rusty steel is best cleaned by sandblasting—a process that will also prepare the concrete surface by removing disintegrated and loose material. The steel should be treated with a rust-inhibitive chemical or coating, which may be a zinc-rich primer or another quality metal primer, and allowed to dry before new concrete is placed.

If the disintegration is deep, it may be necessary to use jackets or forms to hold the fresh concrete in place until it hardens. A bonding agent is recommended to ensure proper adhesion of the new concrete to the base substrate. A latex-cement slurry is the simplest and most economical bonding agent. However, for maximum adhesion, an epoxy bonding agent is recommended. This epoxy compound, painted on the reinforcing steel as well as on the surrounding concrete before placing the concrete, will also serve as a rust-inhibitive primer. A urethane bonding agent may not be suitable since this repair work is not on a deck over occupied areas requiring waterproofing properties.

Repair concrete should have a low water-cement ratio and a low slump. It should also be consolidated in the forms by direct vibration or by vibrating the form. After the concrete has been set for the minimum period of time, the forms may be removed and a liquid sealer may be applied. The sealer will continue to assist in the cure of the concrete and will protect the concrete against weathering.

If the disintegration is shallow, it may be possible to use a latex-fortified or an epoxy mortar to patch the area. The mixtures of latex and cement and sand described under "Repairing Surfaces of Pavement and Slabs" may be used. An epoxy mortar may also be used, but it may not blend readily with the surrounding concrete in color or appearance.

Proper preparation of the surface and proper priming are still necessary. If latex mortar is used, a sealer should be applied. It is not necessary to apply a sealer to an epoxy mortar since this compound is dense and resistant enough to require no additional protection.

Shotcrete or gunned mortar can be used on spalled areas by an experienced contractor who has the necessary equipment. The mortar or cement plaster is usually formulated to a 1:4.5 ratio of cement to sand, although richer 1:3 mixtures are often used. If, as in some cases, the shotcrete does not have the necessary adhesive qualities, a bonding agent (a latex-cement slurry or an epoxy compound) is used as a prime coat. Shotcrete applied in very thin layers may have to be reinforced with a latex emulsion admixture. When the necessary equipment can be obtained, this application can be made by plant maintenance people.

MASONRY*

Masonry structures, like all plant structures, are susceptible to deterioration caused by natural weathering and deleterious effects of the industrial environment. Steps can be taken to modify the rate of attack when the basic principles involved in weathering deterioration are understood.

BRICK MASONRY CONSTRUCTION

Although bricks may be made of many materials, the term *brick masonry* is normally applied only to that type of construction employing comparatively small units of burned clay or shale. Ordinary brick is economical, and, when hard burned and laid in good mortar, it is one of the most durable construction materials available for buildings.

Clay is produced naturally by the weathering of rocks. Shale, produced in much the same way but with compression and perhaps heating, is denser than clay and more difficult to mine. Various brick colors and textures result from different chemical compositions and methods of firing.

Clay is ground, mixed with water, and molded into bricks by several methods: (1) stiff-mud process in which stiff, plastic clay is pushed through a die and cut into desired lengths, (2) soft-mud process in which clay is pressed into forms, and (3) dry process in which relatively dry clay is put into molds and compressed at pressures from 550 to 1500 lb/in^2.

After some drying, green bricks are fired in large kilns. The total firing process takes between 75 and 100 h. The brick must be gradually brought to the vitrification point (the temperature at which clays begin to fuse) and then must be gradually cooled.

Common brick, also known as hard or kiln-run, is made from ordinary clay or shale and is fired in the usual manner. Overburned bricks, called "clinkers," are unusually hard and durable.

A standard brick is 8 in long, 2¼ in deep, and 3¾ to 3⅞ in wide and weighs approximately 4½ lb. It should be rough enough to assure good bonding with mortar and should not absorb more than 10 to 15 percent of its weight in water during a 24-h soaking.

Types of Brick

Common bricks are often classified according to their position in the kiln, as follows:

Arch and *clinker bricks* are close to the fire in the kiln and are overburned and extremely hard and durable. They are often irregular in shape and size.

Red, well-burned and *straight-hard* are well-burned, hard, and durable bricks.

*This material adapted with permission from *Plant Engineering*, November 10, 1977, pp. 203–207.

Stretcher bricks come from these classifications and are selected for uniformity of hardness, size, and durability.

Rough-hard bricks are in the clinker class.

Soft and *salmon* bricks are farthest from the fire in the kiln and are underburned, soft, and less durable.

Special kinds of brick include the following:

Face brick is made from specially selected materials to control color, texture, hardness, uniformity, and strength. It is used in veneering and exterior tiers, chimneys, etc.

Pressed brick is made by the dry process and has regular smooth faces, sharp edges, and perfectly square corners. This type is generally used as face brick.

Glazed brick has the front surface glazed in white or other colors. It is used in dairies, hospitals, and other buildings where cleanliness and ease of cleaning are important.

Fire-brick is made from a special type of fire clay to resist high temperatures.

Imitation brick is usually made from portland cement and sand rather than from clay. It is not burned, but has qualities similar to good mortar.

Mortar

Mortar serves several functions in brick construction. It holds bricks together, compensates for brick irregularities, and distributes load or pressure among the units.

Properties of mortar depend, to a large extent, on the type and quantity of sand used. Good mortar is made from sharp, clean, and well-screened sand. When sand is too fine, the mortar has less "give" and water works out of it, making it stiff and difficult to trowel. It may also set before the bricks can be placed. Too much sand robs the mortar of its cohesive consistency and makes it difficult to work with.

Mortar may be classified into five general types, on the basis of composition: straight lime, straight cement, cement lime, masonry cement, or lime pozzolan.

Natural cement is an important constituent of masonry cement because of its gradual strength-gaining properties, high plasticity, excellent water retention, and good adherence to aggregates. Combining natural cement with portland cement, which has early-strength properties, provides a hydraulic cement ideally suited for masonry mortar. Cement-lime mortars are usually classified in accordance with the ratios of cement, lime, and sand, by volume, in Table 1-2.

TABLE 1-2   Cement Lime Mortars for Masonry Construction

| Mortar | Use | Proportions* | | | Minimum compressive strength, lb/in² |
		Portland cement	Hydrated lime	Damp, loose sand	
Type M	Maximum compressive strength	1	¼	2⅝–3¾	2500
Type S	Maximum bond	1	½	3⅜–4½	1800
Type N	General purpose	1	1	4½–5	750
Type O	Non-load-bearing interior construction	1	2	6¾–9	350

*Parts by volume.

†Sand quantity is 2¼ to 3 times combined volume of cement and lime.

Causes of Deterioration of Brick Masonry

Repeated natural destructive forces, mild though they may be, break down hard rock into clay. These same forces act on clay bricks, fired tile, and fired terra cotta to cause deterioration. Natural stones and the binding mortars of masonry construction are similarly affected. In addition, airborne chemicals in industrial atmospheres and pollutants from internal-combustion engines contribute greatly to rapid soiling and chemical destruction of these binding materials and masonry units.

Frost Damage

One of the more destructive agents of weathering is frost. Water expands 9 percent as it freezes. Under certain conditions, such expansion may produce stresses that disrupt the bricks and cause spalling.

To take up water, a brick must be porous. There are two measures of water absorption: one obtained after soaking the brick for 24 h in cold water and another, larger one obtained after boiling the brick for 5 h. The difference between the two values represents the so-called sealed pores (pores that are not accessible to water under normal conditions, such as wetting by rain). If all open pores are filled with water, the unfilled sealed pores provide space into which water can expand on freezing with little or no development of stress.

Efflorescence

Pitting and spalling of clay products and natural stones is associated with efflorescence, a phenomenon in which salts percolate through the member and crystallize on the surface of a brick, stone, or mortar joint. These salts may be the sulfates of calcium, magnesium, sodium, potassium, and, in some cases, iron. Soluble salts may be present in the clay, or they may be formed by the firing process (oxidation of pyrites, or reaction of sulfurous fuel gases with carbonates in the clay).

Other sources of soluble salts are portland cement and hydraulic lime mortars, which contain soluble sulfates and carbonates of sodium and potassium; mortar containing magnesium lime, which can produce destructive magnesium sulfate; and gypsum plaster or dry wall. Salts may also be drawn by capillary action from the ground soil and limestone or concrete copings.

For efflorescence to occur, salts must be present, water must be available to take them into solution, and a drying surface on which evaporation can proceed to deposit crystals at the surface must exist. Water is always available and drying is periodic. The potential for efflorescence can be reduced by specifying well-fired brick that, generally, contains less soluble salts, or by specifying special-quality brick formulated with minimum soluble salts. The place where efflorescence appears is no indication of its source because solutions of salts may migrate considerable distances. Efflorescence on a particular surface merely indicates that it provided a convenient drying area. When a very dense impermeable mortar touches a more permeable brick, efflorescent salts will often appear on the brick, although the salts may have migrated from the mortar.

Glazed brick veneers do not always eliminate efflorescence. Although the glazing is impermeable to water entry from the exterior, moisture can still move slowly from the interior of the building toward the exterior face. Salts transferred to the exterior glazing create enough pressure to produce shaling of the glazed face. This phenomenon is common, especially when the brick has been inadequately fired. The result is unattractive, and the brick is exposed to further degradation.

Dimensional Changes

Expansion or contraction of building units may not in itself be harmful. However, the continuation of differential movements of dissimilar materials may give rise to difficulties. Live-load changes and foundation settlement can cause whole buildings to move. Building joints must be properly designed and located to accommodate these movements. Lime-sand mortar, used in older structures, is able to accommodate large movements without distress. Modern, higher-strength cement mortars, on the other hand, are less flexible and may shrink excessively on setting to cause cracking.

The movement of moisture in porous building materials can cause expansion and contraction. Expansion generally takes place with wetting; shrinkage occurs with drying.

Water taken up by new brick when it is laid in fresh mortar causes the unit to expand Meanwhile, the mortar is shrinking. Such action can be minimized by wetting kiln-fresh bricks before they are used, avoiding excessively rigid mortars, and providing adequate joints. Reinforced concrete frames may also lead to failure of brick cladding when there is vertical shrinkage of the frame and when no movement joints are provided.

Steel columns clad with brick, common in structures built two decades or more ago, also can cause brick failure. As water enters this cladding, through either the brick or the mortar, the steel column rusts. Expansion of the rust pushes the brick cladding away from the column.

Mortar Deterioration

Mortar may decay from the formation of calcium sulfo-aluminate (which causes expansion and loss of mortar strength) and by the attack of pollutants in the atmosphere. Portland cement contains tricalcium aluminate, which reacts with sulfates in solution to form calcium sulfo-aluminate. Exhaust gases from automobiles contain sulfur dioxide, sulfur trioxide, and nitrous oxides. These oxides react with moisture in the atmosphere to form sulfurous acid, sulfuric acid, and nitric acid, which are the attacking agents. As attack continues over the years, the mortar joints may crack, the surface of the joint may spall off, and the mortar may become softer and more crumbly.

Sulfate-resistant cements used in mortar will inhibit this disintegration.

REPOINTING MORTAR JOINTS

Repointing, or tuckpointing, is the process of removing deteriorated mortar from masonry joints and replacing it with new mortar to correct some perceptible problem, such as falling mortar, loose bricks, or damp walls. All contributing factors to the problem should be thoroughly investigated and corrected before repointing because the great amount of hand work and special materials required make repointing expensive and time consuming. Matching bricks may have to be obtained or specially made. Existing mortar may have to be analyzed before a repointing mortar can be formulated to match its color, texture, and physical properties.

It is a common error to assume that hardness or high strength is a measure of durability. A mortar that is stronger and harder than the masonry units will not "give," causing stress concentrations in the masonry units. These stresses are usually relieved by cracking and spalling. Mortar should contain as much sand as possible (consistent with workability) to help reduce shrinkage while drying. It should have good cohesive and adhesive qualities, be easy to handle on the pointing tool, and have good water retention (to resist rapid loss of water through absorption by the brick). It should not be sticky.

There is some controversy as to whether high-lime mortar is preferable to portland cement mortar for tuck pointing. High-lime mortar is suggested for use on old buildings because it is soft and porous, has low volume change, and is slightly soluble in water.

A slight amount of high-lime mortar will dissolve in rain and precipitate in small cracks; during drying, these small cracks and voids will seal. A small amount of white portland cement will accelerate setting of this normally slow-setting mortar. Even if the building was originally constructed with cement mortar, high-lime mortar may be recommended to reduce shrinkage and potential stresses at the edges of the masonry.

Mortar Mixes

The mixes outlined in Table 1-3 provide a starting point for developing a visually and physically acceptable mortar.

Sometimes, small amounts (5 to 10 percent) of finely divided iron are added to the repointing mixes to provide slight expansion rather than normal shrinkage. Too much iron may produce excessive expansion and may cause iron-rust stain. Repointing mortars

TABLE 1-3   Trial Mortar Mixes for Repointing Masonry

Mortar	Portland cement	Hydrated lime	Sand	Acrylic latex (50% solids)
Formula A	¼ bag	1 bag	3 ft³	—
Formula B	1 bag	1–1½ bags	5–6½ ft³	—
Formula C	1 bag	—	3 ft³	2 gal

can be further modified by adding (1) water-reducing agents to keep water content or water-cement ratio low, (2) waterproofing admixtures such as stearate soaps to minimize the absorption of water, (3) air-entraining agents to increase resistance to freeze-thaw weathering in areas of extreme exposure, and (4) mineral oxide colors to blend the mortar color with the brick.

Requirements for Mortar Materials

Materials used in preparing pointing mortars should comply with the following specifications and requirements:

Lime.   "Standard Specification for Hydrated Lime for Masonry Purposes (ASTM C 207)," Type S; Federal Specification SS-L-351B.

Cement.   "Standard Specification for Portland Cement (ASTM C 150)," Type I or II; Federal Specification SS-C-192G(3). The cement should not have more than 0.60 percent alkali (sodium oxide), or not more than 0.15 percent water-soluble alkali by weight.

Sand.   "Standard Specification for Aggregate for Masonry Mortar (ASTM C 144)," Federal Specification SS-A-281B(1), paragraph 3.1

Water.   Potable water free from acids, alkalis, and organic materials.

Execution of the Work

Generally, old mortar should be cut out to a depth of ¾ to 1 in to ensure an adequate bond between old and new mortar and to prevent popouts. For joints that are less than ⅜ in wide, cutting back ½ in is usually sufficient. Using power tools is risky, unless they are handled by a skilled mason, because bricks can be easily damaged by such equipment. The use of hand chisels is still the best procedure.

Dry mortar ingredients should be mixed first. They should be prehydrated with only enough water to make a damp, stiff mortar to help prevent drying shrinkage. After an hour or two, the mortar is mixed with additional water to provide trowelability.

The joints should be thoroughly cleaned and moistened before the mortar is placed. A chemical bonding agent may be used; however, care must be exercised in painting it into the joint so that it is not smeared on the brick face. Mortar must not be placed before the bonding agent has set.

The mortar is best placed in ¼-in layers and then packed until the void is filled. When the final layer of mortar is thumbprint hard, the joint should be tooled to the desired shape with the correct size of pointing tool. If old bricks have worn, rounded corners, the final mortar surface should be slightly recessed to avoid leaving uneven joints that may be damaged easily.

The small amount of excess mortar left on the wall from a careful repointing job can be removed with a bristle brush before it hardens. Hardened mortar can be removed with a wooden paddle or chisel. Care should be exercised in using any chemicals, especially acids.

Grouting techniques can be used to repoint joints that show only minor defects such as very shallow deterioration, hairline cracks, or slight loss of adhesion to the brick. One method, often referred to as a bagging operation, can be used on glazed brick. A grouting mortar such as formula C in Table 1-3 is mixed to the consistency of a creamy paste and brushed onto the wall surface. After a short period (15 to 30 min, depending on temperature and degree of set) burlap bags or rags are used to bag or wipe the grout from the surface of the nonabsorptive glazed brick; grout is left only in the mortar joint. After an entire section is done, the area may be washed to remove any grout remaining on the glazed brick faces.

Another method involves the same bagging or grouting operation, but it may be done on unglazed or any standard face brick. This method involves masking each brick with tape cut to the same dimensions as the brick. The same type of latex grout is applied by brush over a wider area. The masking tape is removed after the grout sets, leaving neat, clean, and sharp mortar joints. This method is commonly called mask and grout.

Some mortar joints, especially very narrow ones, may be sealed with an elastomeric sealant applied by a caulking gun. Cracked bricks may also be repaired by widening the crack and sealing it with an elastomeric sealant.

CLEANING PLANT BUILDINGS

There are a number of valid reasons for cleaning building exteriors. Before building units are repaired or replaced, the original colors and textures of these units must be known so they can be properly matched. And, when new materials are installed, uniform appearance and weathering of the entire structure are desirable.

Preventive maintenance is an often-overlooked reason for cleaning. Dirt provides a much greater surface area than clean building materials; and, the more surface area that is exposed to atmospheric pollutants, the greater are the possibilities for destructive chemical reactions to be started. Dirty areas remain wet longer, resulting in more severe freeze-thaw cycling. And wet, dirty areas can support microorganisms that can cause disintegration, dissolution, and staining.

Selecting an appropriate cleaning method can be challenging, because dirt composition is so complex. Dirt is a surface deposit of finely divided solids held together by various organic materials. The solids are primarily carbon soot, siliceous dust, and inorganic sulfates. The organic binders consist largely of hydrocarbons from incomplete combustion products of various fuels. A combination of adsorption and electrostatic attractive forces hold the dirt to the masonry. Other adherent factors include efflorescent salts, leached cementitious materials, and recrystalized carbonates.

Acidic cleaners can be very damaging, particularly to marble and limestone, and alkaline cleaners can also be harmful. It is recommended that cleaners be tested on small areas to determine their effectiveness and reaction to the substrates.

One of the most versatile techniques for cleaning building exteriors is water washing. Although it requires minimal expenditure for materials and equipment, it can be time consuming, particularly if hand scrubbing is involved. There are three types of water-washing procedures: low pressure, high pressure, and steam cleaning.

The low-pressure wash is carried out over an extended period. The prolonged spraying loosens heavy dirt deposits; then, moderate pressure (200 to 600 lb/in²) can be used to flush away loosened dirt.

High-pressure water cleaning involves equipment capable of supplying water to a special high-pressure gun that jets the water at pressures of 1000 to 1800 lb/in². The gun can deliver up to 1700 gal of water per hour. A special nozzle can aerate the water, thereby minimizing physical damage to the masonry.

Steam cleaning involves the use of low-pressure (10 to 30 lb/in²), large-diameter (½ in) nozzles; steam is generated from a flash boiler. The equipment is relatively more expensive and presents some safety hazards to the operators. It is also possible to add detergents or surfactants to the water in the flash boiler. Adequate dirt removal requires an average working time of 1 min/ft².

Chemical cleaners can be acidic (low pH) or alkaline (high pH). Both types are used with surfactants (1 to 2 percent) to promote detergency and wetting. Acidic cleaners are often based on hydrofluoric acid or phosphoric acid in concentrations of less than 5 percent in water. Alkaline cleaners are often based on sodium hydroxide, ammonium hydroxide, or ammonia. Sometimes, all purpose cleaners, such as ammonium bifluoride, are used.

Abrasive blasting, both dry and wet processes, are effective on all substrates, but they require experienced mechanics to minimize damage to surfaces. Pressures used are usually between 20 and 110 lb/in^2, and working distances are from 3 to 12 in. The abrasives are usually silica sand, but crushed slags and coal wastes are often used. The mesh sizes are very fine, either 0 or 00. Round particles are less abrasive and damaging than crushed grains.

Cleaning may precede or follow replacement of masonry units and repointing, depending on conditions. Cleaning before repair helps reveal original colors and textures and prepares substrates for receiving new mortars and bonding agents. Cleaning after repairs helps remove any excess droppings, splashings, or other accidental contamination. Whether done before or after, cleaning is a very important aspect of masonry and concrete restoration.

SEALING THE SURFACE

A final step in the entire restoration process involves sealing all porous surfaces on the exterior facade. This procedure waterproofs the masonry to minimize the ingress of water, protects against attack by pollutants and other chemicals, minimizes the collection of dirt, and protects against graffiti damage.

One of the best sealers is a methyl methacrylate in organic solvent, containing 15 to 20 percent solids and, preferably, having a matte finish. The sealer may be applied in one or two coats, depending on the porosity of the substrate, by brush, spray, or roller. The sealer is water white, will not yellow or embrittle, and may be effective for 5 to 10 years or longer.

Silicone compounds are not particularly recommended, primarily because they are effective only for relatively short periods and are water-repellent rather than waterproofing.

METALS*

GLOSSARY OF COMMON METAL CORROSION TERMS

Active   A state in which a metal tends to corrode (opposite of passive, noble).

Anode   Electrode in a cell at which surface oxidation or corrosion occurs (opposite of cathode).

Bimetallic corrosion   Galvanic corrosion.

Cathode   Electrode in a cell at which reduction occurs (opposite of anode). Practically no corrosion occurs here.

Cell   An electrochemical circuit consisting of an anode, a cathode, and electrolyte. The anode and cathode may be different metals or dissimilar areas of same metal. Corrosion generally occurs only at anodic areas.

Corrosion   Direct, chemical or electrochemical reaction of a metal with its environment, and general destruction of any material resulting from reaction with environment.

Corrosion fatigue   Damage to metal from combination of corrosion and fatigue (cyclic stresses).

Couple   A cell formed by two dissimilar metals in an electrolyte.

Electrolyte   A chemical mixture capable of carrying ions. It is usually a liquid (often an aqueous solution).

Erosion   Deterioration of a surface by the abrasive action of moving fluids.

*This material adapted with permission from *Plant Engineering*, November 24, 1977, pp. 103–106.

Erosion-corrosion Combined effects of erosion and corrosion on a metal surface.

Fatigue Cracking failure of a material resulting from repeated cyclic stress below the normal tensile strength.

Galvanic cell An electrolytic cell, consisting of two dissimilar electrodes (anode and cathode) in an electrolyte, capable of producing electric energy by electrochemical action. The anode and cathode may be dissimilar metals or dissimilar areas of the same metal. Corrosion occurs at anodic areas.

Galvanic corrosion Corrosion associated with the current in a galvanic cell (also called couple action).

Galvanic series Listing of metals arranged according to their relative corrosion potentials in some specific environment (often seawater).

Ion An electrically charged atom or group of atoms (radical).

Local cell Galvanic cell produced by differences in composition in the metal or the electrolyte.

Noble metal A metal, or alloy such as silver, gold, copper, having high resistance to corrosion or oxidation.

Oxidation Loss of electrons by a chemical reaction such as corrosion (opposite of reduction).

Passive A state of a metal that tends to slow corrosion or anodic reactions (opposite of active). It is a surface phenomenon.

Patina A green coating that slowly forms on copper and copper alloys exposed to the atmosphere. It consists mainly of copper sulfates, carbonates, and chlorides.

Reduction Gain of electrons (opposite of oxidation).

Rust The reddish-brown corrosion product of iron and ferrous alloys. It is primarily hydrated ferric oxide.

Corrosion is the critical performance factor for metals in structures exposed to weather and industrial atmospheres. Corrosion can be completely destructive, or it can form a protective film over the surface, stopping the corrosion process. Methods of controlling the destructive effects of corrosion vary according to the type of metal and the environment.

Corrosion may be defined as the direct chemical or electrochemical reaction of a metal with its environment. Chemical attack is simple dissolution of the metal. Electrochemical attack requires an electrolyte, anode, cathode, and return circuit (see Fig. 1-2). The electrolyte of a simple galvanic corrosion cell is a substance capable of carrying ions. It is typically an aqueous solution of small amounts of dirt, salts, acids, or alkalis from the atmosphere and rain or condensed moisture. The anode and cathode of the corrosion cell may be dissimilar metals or two areas of the same metal in dissimilar electrolytes.

The corrosion potential of a galvanic couple (galvanic corrosion cell of dissimilar metals) is indicated by the position of the metals in the galvanic series—a list of metals and alloys arranged according to their relative corrosion potentials—for the given environment (seawater in the example shown in Fig. 1-3).

Most metals used in building construc-

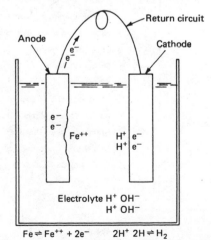

Figure 1-2 Four essential elements of the basic corrosion cell are electrolyte, anode, cathode, and return circuit.

Active

↑

Magnesium

Zinc

Beryllium

Aluminum Alloys

Cadmium

Mild Steel, Cast Iron

Low Alloy Steel

Stainless Steel (Active) Types 410, 416, 430, 302, 304, 321, 347

Austenitic Nickel Cast Iron

Stainless Steel (Active) Types 316, 317

Aluminum Bronze

Naval Brass, Yellow Brass, Red Brass

Tin

Copper

Pb-Sn Solder (50/50)

Admiralty Brass, Aluminum Brass

Manganese Bronze

Silicon Bronze

Tin Bronzes (G & M)

Stainless Steel (Passive) Types 410, 416

Nickel Silver

90-10 Copper-Nickel

80-20 Copper Nickel

Stainless Steel (Passive) Type 430

Lead

70-30 Copper-Nickel

Nickel-Aluminum Bronze

Nickel-Chromium alloy 600

Silver Braze Alloys

Nickel 200

Silver

Stainless Steel (Passive) Types 302, 304, 321, 347

Nickel-Copper alloys 400, K-500

Stainless Steel (Passive) Types 316, 317

Nickel-Iron-Chromium alloy 825

Ni-Cr-Mo-Cu-Si alloy B

Titanium

Ni-Cr-Mo alloy C

Platinum

↓

Graphite

Noble or Passive

tion are alloys: single-phase alloys in which the atoms of the different metals are combined in a single crystal structure, or two-phase or multiphase alloys in which the metals are physically distinct. Because separate phases can form anodic and cathodic regions, a multiphase alloy is more susceptible to certain types of galvanic corrosion than a single-phase alloy or a pure metal. Deformation of stress of a metal member can also produce galvanic cells that lead to corrosion.

Many metals used in buildings become passive when exposed to the atmosphere. Such metals include aluminum, stainless steel, copper, lead, and zinc. They corrode on initial exposure to the atmosphere and then acquire a protective layer of corrosion products that isolates the metal from further attack. The green patina of copper is one of the most visible examples of this phenomenon.

Corrosion is promoted by a variety of conditions—contact of metals with other building materials (such as uncured concrete, mortar, or plaster), salts in the soil, contaminated rainwater, or acids from certain types of timber. Calcium chloride added to concrete during cold-weather construction can intensify galvanic corrosion of embedded metals.

ALUMINUM

The four principal categories of wrought and cast aluminum for construction are various grades of pure aluminum, heat-treatable alloys, non-heat-treatable alloys, and casting alloys. Pure aluminum, which is about as ductile as lead, is used for supported roofing and flashing. Heat-treatable alloys, based on blends of aluminum, magnesium, silicon, and copper, are used for fastenings and structural purposes because of their high strengths. Non-heat-treatable alloys containing manganese or magnesium are used in sheet roofing and cladding. Casting alloys usually contain up to 12 percent silicon and perhaps some copper and magnesium.

Aluminum may be used as manufactured, or it may be furnished with a wide range of treatments. The best-known treatment is anodizing, in which the thickness of the surface film is increased by electrolytic oxidation. Because the film is a suitable base for dyeing, organic dyes are included in the electrolytic bath to produce a wide range of colors in the finished aluminum. The anodic film should be 1 to 2.5 mils thick if it is to be exposed to the weather.

Aluminum may also be surface-treated with chromates, phosphates, or fluorides. Vitreous enameling and various types of paints are also common treatments. Manufacturer-applied finishes, usually baked polymeric systems, are popular for roofing and siding applications.

Prolonged exposure to the atmosphere, particularly in areas of high pollution, causes spots of white crystalline corrosion products to form on the surface of aluminum, including the anodized type. In a sooty atmosphere, the metal may acquire a gray color. Pitting corrosion can affect the useful life of aluminum members in industrial atmospheres.

Generally, maintenance of aluminum building members consists of washing the surface fairly regularly to remove foreign matter that can damage anodized finishes. Washing is particularly recommended for areas that are not effectively washed by rainfall. Other finishes also benefit from washing.

Aluminum members are often shop-coated with a 1- to 2-mil-thick film of an acrylic sealer. This film will ultimately weather to a point at which it must be renewed if protection is to continue. A methyl methacrylate sealer, similar to those used in sealing con-

---

Figure 1-3 (facing page) Galvanic series of metals and alloys in seawater. Metals and alloys are listed in the order of their corrosion potential in seawater. When two metals are coupled, those close together in the list are less susceptible to galvanic corrosion than those widely separated in list.

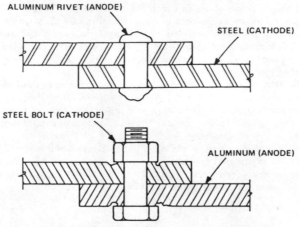

Figure 1-4   In a galvanic couple of aluminum and steel, anodic aluminum corrodes while cathodic steel is relatively unaffected. It is important that fasteners be compatible with building materials.

crete and masonry structures, may be used. Painted aluminum may be touched up or repainted.

For structural applications, it is suggested that:

- Aluminum roofs be pitched to prevent water ponding and promote washing of the surface by rain
- Vapor barriers and proper ventilation be used to minimize condensation on the underside of an aluminum member
- Aluminum fittings (or nonmagnetic stainless steel and galvanized fittings) be used with aluminum building components to prevent galvanic corrosion (see Fig. 1-4)
- Aluminum and steel be separated by insulating washers, bituminous or plastic materials, or paint
- Water from copper roofs or pipes be prevented from discharging onto aluminum
- Aluminum embedded in concrete, masonry, or stonework be protected with a bituminous paint
- Aluminum that will contact certain timbers be protected by paint or bituminous coating to reduce the chance of attack from acidic wood (unseasoned oak, western red cedar, and redwood) and wood preservatives that contain water-soluble salts or copper salts that can corrode aluminum.

ZINC

Although zinc can be used in the form of rolled sheet and strip, its widest use in building is for protective plating. When exposed to weather, zinc on the surface of galvanized steel develops a passive film that protects the underlying zinc and, thus, the steel. When the surface is scratched or scored, the anodic zinc corrodes, protecting the cathodic steel from attack.

Zinc can be applied to metal surfaces with hot-dip galvanizing, electrolytic processes, metal spraying, sherardizing, and zinc-rich paints. Hot-dip galvanizing provides excellent coverage and bond, leaving a 3- to 5-mil-thick film. The interface is actually a zinc-iron alloy. Electrolytic processes provide a thin coating up to 1 mil thick. Metal spraying provides a 4- to 20-mil-thick zinc film. Sherardizing, the heating of small components in a container of zinc dust, produces films 0.5 to 1.5 mils thick. The dry film of zinc-rich paint should be at least 90 percent zinc to provide adequate galvanic protection to a steel substrate.

Atmospheric exposure gives the zinc surface a coating of zinc carbonate that reacts with sulfurous and sulfuric acids in the atmosphere to form zinc sulfate, which is water-soluble and can be washed off by rain.

Zinc sheeting is susceptible to attack from condensation on its underside. Proper ventilation of roof spaces and using vapor barriers and suitable underlays can minimize such conditions. Zinc surfaces should be designed to allow rapid drying.

Local air pollution should be considered when designing zinc roofs, and proper fasteners should be specified. Zinc should be prevented from contacting copper and the chlorides and sulfates in concrete and mortar. Acids from certain timbers and some wood preservatives can also attack zinc. Bituminous coatings are recommended for protection.

Weathered galvanized members can be painted after being properly cleaned. Freshly galvanized members can also be painted. The Zinc Institute now recommends solvent cleaning or detergent washing, rather than acid etching, before painting. Suitable primers include zinc dust, zinc oxide primers, vinyl wash primers, and latex emulsion primers.

COPPER

One of the major uses for copper in buildings is in roofing—gutters, flashing, and cladding. It is also used as an alloying metal in brass, bronze, and weathering steel.

Characteristically, copper weathers by the formation of a protective green patina, which, chemically, consists of copper hydroxide salts of sulfate, chloride, nitrate, and carbonate. This process may take 5 to 10 years, depending on air pollution, humidity, and temperature. The patina can be produced artificially, so new copper can be made to match old copper.

Protective lacquers are available that preserve the polished finish of copper or brass. One such coating is Incralac A (a trademark of the International Copper Research Association). It is a methyl methacrylate resin dissolved in a blend of xylol and toluol with a thixotropic agent to leave thicker films, and benzo-triazole, an inhibitor that prevents discoloration of the metal.

Copper shows good compatibility with other building materials. It has long been used with concrete as a water-stop material. Its use in roofing is traditional because of its outstanding durability.

LEAD

In its pure form, lead has been used in plumbing, roofing, cladding, and waterproofing applications. However, lead is now typically alloyed with 6 to 7 percent antimony to increase stiffness and strength. It can be used to coat iron, steel, and copper.

Lead's shiny metallic appearance dulls rapidly in the atmosphere through the formation of an oxide; with prolonged exposure, carbonates and sulfates form, leaving a whitish-gray appearance. Because lead is a relatively soft sheet material and has little rigidity, it requires strong supporting structures.

Lead's galvanic action is relatively insignificant, but contact between lead and aluminum should be prevented. Lead can corrode in contact with damp, uncured, cementitious materials and damp acidic timbers. The corrosion is minor and can be prevented if the lead is coated with bituminous paint. However, most paints do not adhere well to lead.

STEEL

Although there are many uses for wrought and cast iron in building construction, steel is the major type of ferrous metal used in plant construction. Steel and iron deteriorate primarily from four basic causes.

- Corrosion (rusting) is the chief cause of damage to steel and creates a major maintenance problem. The corrosion products of steel are reddish brown scales or flakes of iron oxides, hydroxides, and other compounds loosely bonded to the parent metal. Rusting is an electrochemical process that requires an electrolyte (usually water).

- Abrasion (erosion) may be distinguished by the worn, smooth appearance of the surface. It usually results from the working of moving parts, members subjected to wave action by water (such as piling), flue gases with a high ash content, and windborne dust, sand, and debris.

- Fatigue is failure of a structural member resulting from repetitive, fluctuating loads at or below allowable design values. Symptoms are small, often undetectable, fractures which may result in sudden collapse. Steel members subjected to repetitive loadings should be inspected frequently, particularly at riveted and welded points.

- Impact of moving objects can lead to distortion. Sometimes, bowing or buckling can result from overstress, weakening of a steel member from corrosion, or both. Corrosion of steel lintels over windows can reduce effective thickness, resulting in bowing from the weight of bricks above the lintel.

Corrosion and abrasion problems can be solved by proper maintenance. Impact and fatigue are engineering considerations.

Corrosion of iron and of carbon and low-alloy steels can be controlled by protective coatings. Corroded members must be properly cleaned before appropriate primers and finish paint systems are applied. Abrasive blasting is the most effective method of preparing corroded steel for painting. Cleaning to commercial grade is usually adequate, but blasting to white metal may be necessary when high-performance coatings such as epoxies or urethanes are to be used, or when the steel is to be exposed to a highly corrosive atmosphere. Other cleaning methods include power wire brushing, hand wire brushing and scraping, and flame cleaning.

STAINLESS STEEL

High strength and outstanding resistance to atmospheric corrosion are major advantages of stainless steels (alloys containing from 17 to 20 percent chromium). Various modifications of these alloys incorporate up to 8 percent nickel and 3 percent molybdenum. The straight chromium alloys are magnetic; those containing nickel and molybdenum are nonmagnetic. Stainless steel is used in window framing, curtain walls, cladding, and fasteners.

A stainless steel surface develops a thin oxide film that is highly corrosion-resistant and self-repairing. Regular washing to remove dirt and other deposits is normally the only maintenance required. Stainless steel causes very minimal galvanic corrosion of other metals.

Weathering steel, a high-strength, low-alloy steel for architectural applications, also develops a passive surface. These alloys, containing small percentages of carbon, manganese, phosphorus, sulfur, silicon, copper, chromium, and nickel, develop a deeply colored, reddish-brown coating. It develops with time and only with complete exposure to the elements. Until the weathered surface comes to equilibrium and the coating develops, adjacent concrete, unglazed brick, stucco, granite, marble, unpainted galvanized steel, flat and semigloss paints, and some types of glass may develop a rust stain. Unexposed areas of this steel, like ordinary carbon steel, must be painted.

WOOD AND PLASTIC*

Wood and plastic building materials resist many of the natural and industrial environmental factors that affect concrete, masonry, and metals. However, wood and plastic are

*This material adapted with permission from *Plant Engineering*, February 16, 1978, pp. 149–152.

susceptible to other types of attack—fungi and insects can destroy wood, and ultraviolet light and temperature can affect some plastics.

WOOD

When exposed to the elements, wood is affected by water and light. Pollutants have little effect on wood except to dirty it. It has good resistance to chemical corrosion, and its thermal expansion is slight.

Wood normally contains 12 to 18% water, but it can absorb moisture and swell up to 5 percent. Ultraviolet light can cause chemical and color change (accelerated by the extraction of water-soluble materials). The stresses set up by fluctuating moisture content and ultraviolet light cause splitting and checking (defects in the wood surface that allow more water to enter to produce gross dimensional changes). Weathering effects of moisture and sunlight can be reduced by applying a film-forming or penetrating finish.

Biological Attack

Wood is subject to decay caused by certain types of fungi. In extreme cases, such as in tropical climates, unprotected timber may be destroyed in a few months. Keeping wood adequately dry (less than 20 percent moisture) minimizes fungal damage. Structural timber should be protected from the weather by a well-ventilated shelter, and wood members should not be framed close to the ground.

Deprivation of air causes fungi to become dormant and, finally, die. Encasement in concrete or heavy bituminous coatings can shut off the air supply; ordinary painting will not. Embedment of timbers in soil can also shut off access to air, but it can cause other problems.

Temperatures between 50 and 90°F are optimum for the growth of fungi. They become dormant at temperatures over 110°F and near 32°F, but can reactivate when temperatures moderate.

Protection against fungal decay can be provided by various chemical treatments.

In wood exposed to marine atmospheres, attack by marine borers may lead to destruction. Aggressive organisms such as teredo or shipworms, various mollusks, and wood lice attack wood by burrowing or boring and gnawing.

Their terrestrial counterparts—termites, beetles, caterpillars, bees, and ants—eat the wood. Such attack reduces the strength of the timbers and weakens and ultimately destroys the structure.

Termites are one of the most destructive organisms to timber. They are found in almost all areas of the country, with greater concentrations in warmer and more humid regions. Their diet is based on cellulose. There are wood-dwelling and earth-dwelling types. Visible symptoms of termite attack are subtle and difficult to recognize.

Protection of most wood against termites is provided by chemical treatments or by poisoning the ground with chemicals such as copper sulfate, sodium fluosilicate, borax, paradichlorobenzene, and various commercial poisons.

Chemical Preservatives

A number of species of wood have natural decay resistance. The heartwood of several species native to North America—especially redwood, cedar, cypress, and juniper—has good decay resistance; however, the sapwood of substantially all common species has poor resistance. Regardless of the species, chemical protection of woods is recommended. The preservative treatment makes the wood poisonous to fungi, insects, and marine borers. There are three classes of preservatives: waterborne, oilborne, and creosote.

Waterborne Salts

Waterborne salts for wood preservation include zinc chloride, chromated zinc chloride, copperized chromated zinc chloride, zinc meta arsenite, chromated zinc arsenate, chromated copper arsenate, ammoniacal copper arsenite, acid copper chromate, and fluor chrome arsenate phenol. These chemicals leave little odor and have little effect on the appearance of the wood, which may also be painted. The zinc chloride salts, at high penetrations, also provide fire retardance. Wood should be reseasoned before use because this type of chemical treatment injects a large amount of water into the wood.

Oilborne Preservatives

These include penta (pentachlorophenol) and copper naphthenate. Penta does not change the color of the wood, but copper naphthenate gives wood a green shade. Paintability after treatment depends on the type of oil or solvent used as a vehicle.

Creosote

Creosote is excellent for preventing decay, especially in exterior use or in contact with water. Its advantages include relative insolubility in water (giving it a high degree of permanence), good penetration, good availability, and low cost; in addition, it causes little dimensional change in the wood. Disadvantages include its potential fire hazard; a distinctive, sometimes unpleasant odor that can affect foods; volatile vapors that can affect plant life; black color; and staining of adjacent woods or porous materials. Treated wood cannot be painted, and on hot days it may sweat and become wet and tacky.

Treatment Processes

Methods for treating timber with chemicals include pressure, coating, dipping and steeping, thermal, and diffusion.

Pressure Process

These processes produce relatively deep penetration of the preservative into the wood. Although various processes differ in detail, the basic principle of all is the same—wood is placed in a pressure vessel that is filled with preservative. Pressurization then drives preservative into the wood to meet penetration and retention specifications.

Coating

Coating by brush or spray may be done at the site. At least two, preferably three, applications are made; each coat is applied after the previous one has been absorbed. Brush and spray treatments are generally used only when more effective treatments are not practical.

Dipping and Steeping

This involves immersing the wooden member in the preservative liquid. From a few minutes to as long as several days of immersion may be required. It provides greater penetration of the preservative than brush and spray treatments, but it generally less effective than pressure processes.

Thermal Treatment

Thermal treatment, or hot and cold dipping, is similar to dipping, except the member is first heated in the preservative in an open tank and is then submerged in cold preservative. This procedure provides deeper penetration of preservative than dipping and steeping.

Diffusion Processes

Diffusion processes can be used while the timber is in place. Water-soluble preservatives, carried in bandages or pastes or in retaining rings applied to the member, diffuse into the water present in the wood.

All of these methods should meet the standards and specifications of the American Wood-Preservers' Association (AWPA); see Tables 1-4 to 1-6.

TABLE 1-4   Selected AWPA Standards for Preservatives

Standard numbers	Preservative	Symbol	Trade names†
P1 or P13	Coal tar creosote	—	—
P2 or P12	Creosote–coal tar solution	—	—
P5	Acid copper chromate	ACC	Celcure*
	Ammoniacal copper arsenite	ACA	Chemonite*
	Chromated copper arsenate	CCA	Type A: Greensalt*
			Langwood*
			Type B: Boliden* CCA
			Koppers CCA-B
			Osmose K-33*
			Type C: Chrom-Ar-Cu(CAC)*
			Osmose K-33C*
			Wolman* CCA
			Wolmanac* CCA
	Chromated zinc chloride	CZC	
	Fluor chromate arsenate phenol	FCAP	Osmosalts* (Osmosar*)
			Tanalith*
			Wolman* FCAP
			Wolman* FMP
P8 and P9	Pentachlorophenol	Penta	—

*Reg. U.S. Pat. Off.
†*Source:* American Wood-Preservers' Association.

TABLE 1-5   Selected AWPA Standards for Pressure-Treatment Processes

AWPA* standard	Product
C1	All timber products (general)
C2	Lumber, timber, bridge ties, and mine ties
C3	Piles
C4	Poles
C9	Plywood
C11	Wood blocks for floors and platforms
C14	Wood for highway construction
C18	Piles and timbers for marine construction
C23	Round poles and posts used in building construction
C29	Lumber to be used for the harvesting, storage, and transportation of foodstuffs

*American Wood-Preservers' Association.

Repairing Termite-Damaged Wood

Wood that has been weakened or damaged by termite attack may be repaired without removing the wooden member from the structure. The procedure is similar to pressure-grouting cracks in concrete. For termite-damaged wood, a form sleeve must be built around the damaged wooden member; then, a low-viscosity epoxy-resin compound is injected until all voids have been filled. The form sleeve may be treated with oil or wax to prevent the epoxy resin from sticking to the form. After the epoxy resin has hardened, the wood member's structural strength may be even greater than originally. If, as so often happens, the termite damage is inside the wooden member and the outside shell is intact, the epoxy-resin sealer may be invisible. However, should it be exposed, its light amber color may blend with the wood's color.

Sometimes, epoxy-resin compounds used for repairing wood may be manufactured with protective chemicals such as pentachlorophenol to help prevent further attack by termites. Termites cannot damage the epoxy compound. These special epoxy-resin compounds are generally available in boat yards or marinas, where they are widely used to repair wood damaged by rot or marine borers.

TABLE 1-6 Preservatives and Minimum Retentions for Various Wood Products[a]

Product and service condition	AWPA product[b] standard	Recommended minimum net retention, lb/ft³[c]						
		Waterborne preservatives[d]					Oilborne[d,e]	
		CCA	ACA	ACC	CZC	FCAP	Penta[f]	Creosote
LUMBER AND TIMBER								
Above ground	C2	0.23	0.23	0.25	0.46	0.22	0.40	8
Soil or water contact								
Nonstructural	C2	0.40	0.40	0.50	NR	NR	0.50	10
Structural	C14	0.60	0.60	NR	NR	NR	NR	12
In salt water	C14	2.5	2.5	NR	NR	NR	NR	25
PLYWOOD								
Above ground	C9	0.23	0.23	0.25	0.46	0.22	0.40	8
Soil or water contact	C9	0.40	0.40	0.50	NR	NR	0.50	10
PILING								
Soil or fresh water	C3	0.80	0.80	NR	NR	NR	0.60	12
In salt water								
Severe borer hazard (Limnoria)	C18	2.5 & 1.5[g]	2.5 & 1.5[g]	NR	NR	NR	NR	NR
Moderate borer hazard (Pholads)	C18	NR	NR	NR	NR	NR	NR	20
Dual treatment (Limnoria and pholads)								
First treatment	C18	1.0	1.0	NR	NR	NR	NR	NR
Second treatment	C18	—	—					20
POLES								
Utility in normal service	C4	0.60	0.60	NR	NR	NR	0.38	7.5
Utility in severe decay & termite areas	C4	0.60	0.60	NR	NR	NR	0.45	9.0
Building poles (structural)	C23	0.60	0.60	NR	NR	NR	0.45	9.0
POSTS								
Fence								
Round, half-round, quarter-round	C14	0.40	0.40	0.50	NR	NR	0.40	8
Sawn four sides	C14	0.50	0.50	0.62	NR	NR	0.50	10
Guardrail and sign								
Round	C14	0.50	0.50	NR	NR	NR	0.50	10
Sawn four sides	C14	0.60	0.60	NR	NR	NR	0.60	12

[a] Key to symbols: ACA, ammoniacal copper arsenate; ACC, Acid copper chromate; CCA, chromated copper arsenate; CZC, chromated zinc chloride; FCAP, Fluor chrome arsenate phenol; NR, not recommended; Penta, pentachlorophenol.

[b] See Table 1-5.

[c] Minimum net retentions conforming to AWPA standards for softwood lumber and plywood. Retentions for piles, poles, and posts are for southern pine. AWPA Standard C1 applies to all processes.

[d] See Table 1-4.

[e] Creosote, creosote–coal tar solution, and oilborne penta are not recommended for applications that require clean, paintable, or odor-free wood.

[f] Penta can be applied in liquid petroleum gas or light solvents to provide a clean, paintable surface

[g] Two assay zones: 0 to 0.5 in and 0.5 to 2.0 in.

PLASTICS

Use of reinforced plastics in all construction is growing by millions of pounds yearly. These materials are finding extensive use in ceiling and floor systems, piping, skylights, translucent panels, structural shapes, grating, etc.

Plastics are usually strong, durable, lightweight, and resilient; they are easy to manufacture and install and have low maintenance costs. Color can be built in over a wide range, enhancing their use for exterior decoration.

Plastics, a term used for synthetic or modified natural polymers, are classified into two major groups: thermoplastics, which can be softened by heat and lend themselves readily to molding, and thermosetting plastics, which do not soften under heat once they are cured. Plastics used for exterior applications in the construction industry can be formulated to almost any set of properties, including light stability, rigidity, opacity, water absorption, and abrasion and fire resistance.

Weathering of plastics involves ultraviolet radiation, infrared radiation, water, temperature, microorganisms, industrial gases, and stresses from wind and snow loadings. Some plastics discolor after weathering, but discoloration can usually be minimized by selecting materials containing ultraviolet absorbers. Certain pigments can also increase stability for exterior exposure.

Thermoplastic polymers should not be used in areas where high temperatures are present to cause distortion. Distortion temperatures of commercial plastic materials should be a guide in this respect. Low temperatures can cause embrittlement. Oxidation can cause changes in the molecular structure of a plastic similar to those in paint films. In most cases, plastics are reasonably resistant to industrial pollution and microorganisms, but they will discolor as dirt collects. However, frequent simple washings can maintain plastics adequately.

The most important generic types of plastics for exterior application include polyvinyl chloride, glass-reinforced polyester, acrylics, phenolics, and amino resins.

Polyvinyl Chloride (PVC)

One of the most extensively used plastics in construction, PVC is often used for roofing panels, gutters and downspouts, pipes, cladding, wall and floor coverings, and window frames. PVC film has also been used extensively on metal sheeting as a protective and decorative finish. Heavy use is made of PVC in hidden items, such as water stops, vapor barriers, and waterproofing membranes.

PVC is a thermoplastic material with a wide range of properties determined by stabilizers, plasticizers, ultraviolet absorbers, lubricants, and other additives. For fire-resistant properties, antimony trioxide is often used. Methods of manufacture include extrusion, injection molding, blow molding, and calendering.

When properly formulated and fabricated under controlled conditions, a PVC product can have a life of 20 to 30 years. Certainly, a PVC waterstop, buried in a concrete foundation wall, must last for the life of the structure.

Translucent or transparent PVC sheet does not weather as well as the opaque form because stabilizers and ultraviolet absorbers affect light-transfer properties.

If weathering causes color change or differential fading, the PVC member can be painted after proper preparation of the surface by washing and light sanding.

Glass-Reinforced Polyester

Laminates of glass-reinforced polyester find wide use in automobile bodies, aircraft, boats, swimming pools, tanks, prefabricated housing systems, curtain walls, and lightweight building panels.

Polyesters are thermosetting resins produced by the reaction of mixtures of glycols and dibasic acids. The compound is comparable to an alkyd resin used in paints. However, it is further modified by dissolving it in styrene, and it is cured into a thermosetting plastic by the addition of catalysts and accelerators. The plastic can be modified with additives similar to those used in the PVC compounds.

The polyester resin, by itself, is not strong enough for industrial use, so it must be reinforced with glass fibers. A laminate is manufactured by spreading the catalyzed polyester resin onto a form or mold. While it is still uncured, woven glass cloth, swirled mat, or chopped strands are laid up into the film. Then, more coats of catalyzed polyester and glass fibers are added until the necessary number of layers is installed. The final coat of polyester is normally heavy enough to cover the glass fibers thoroughly. The last layer of resin will be exposed to weathering and can have color and all the necessary additives built in.

The polyester resin may be modified with methyl methacrylate resins for better clarity and transparency. Recent developments involve the use of acrylates or polyvinyl fluoride as surface coatings bonded to the laminate for longer gloss retention. The product is cured at ambient temperatures, although heat will accelerate the cure and ensure a more satisfactory laminate. The same procedures are also followed in producing epoxy-resin laminates, although these are not as widely used in the construction industry.

A glass-reinforced polyester laminate may fail if the surface resin is not thick enough to cover the glass strands, or if weathering wears away the surface color. Change of color, fading, and loss of gloss may also develop. Good maintenance depends on regular and thorough washing to remove dirt. When the surface must be refurbished, steel wool may be used to remove dirt and loose resin and fibers. A surface layer of polyester resin or an acrylic sealer may be applied. Frequent washing to remove dirt is usually the best method of maintenance.

Acrylics

Methyl methacrylate is the basic monomer for acrylic resins, materials that found their first extensive uses in cockpit covers of airplanes. Today, methyl methacrylate and its modifications are used in making window panes, fascia panels, skylights, sunshades, bath and shower enclosures, roof lights, etc. These thermoplastic materials are water-white and have almost the same light transmission properties as glass. They have good impact resistance, can be easily formed and machined, are easy to handle and install, and possess outstanding weathering resistance and durability. However, acrylics have low abrasion resistance and a very high coefficient of thermal expansion.

Periodic cleaning and washing of acrylic members is recommended as the best method of maintenance. However, they are easily scratched. Polishing with a soft rouge can remove scratches without affecting transparency. Colored acrylic members may be produced, and there is little or no fading or change in color. Painting is seldom required.

Phenolics and Amino Resins

Reacting aldehydes with phenol and amino compounds (such as urea and melamine) can produce thermosetting plastics with good chemical resistance. One of the first synthetic resins of this type was Bakelite, a phenolic resin. This group of plastics is used in making laminates, usually with reinforcement, for curtain wall paneling, wall linings, corrugated roofing, etc. Melamine formaldehyde is used in making the laminate known as Formica. These plastics, particularly the phenolics, can change color and weathering and develop very fine crazing patterns with aging. Painting can restore surface appearance and color after proper washing and light sanding. Frequent washing is recommended for maintenance.

Other plastics are used in plant structures, although less extensively. Among these are the acrylonitrile butadiene styrene (ABS) resins, the polyvinyl fluoride resins, the polycarbonate resins, and the polyurethanes. The epoxy resins have been used extensively in structural applications (such as flooring) and adhesives. Specialized applications include use in chemically resistant coatings and in plasters for exposed aggregate wall finishes.

As research and development proceed, improved formulations of available plastics and completely new resins will find their way into plant structures.

chapter 3-2

Paints and Protective Coatings

by
Robert J. Klepser
Senior Development Chemist
Heavy Duty Maintenance Coatings
PPG Industries, Inc.
Coatings and Resins Division
Houston, Texas

BASICS OF PROTECTIVE COATINGS

The Corrosion Process

This chapter covers the use of coatings to protect a variety of surfaces. A discussion of corrosion and other surface deterioration can be found in Chap. 3-1.

Because steel is most susceptible to corrosion attack, it is, in most situations, the surface of primary concern. A clear understanding of the processes by which a useful steel structure is reduced to a collection of rusty scrap is essential to controlling corrosion and making intelligent recommendations for protective coatings systems.

The reddish rust on a piece of steel is created electrochemically by the iron in it combining with atmospheric oxygen. Most of the iron ores found in nature are oxides, or combinations of iron and oxygen. The most common form of iron ore is hematite (Fe_2O_3), an oxide of iron which is equivalent to the form of common rust. These natural oxides are converted into usable iron and steel products by exposing the ores to a very vigorous reduction reaction which separates the iron and oxygen. Because iron has a strong affinity for oxygen it is necessary to deal with the ever-present tendency of iron to recombine with oxygen and form the stable natural oxides once again. This change is an electrochemical process accompanied by the production of minute, measurable electric currents—a process very similar to that creating electricity in a battery. Therefore, to create rust or iron oxide on a piece of steel, there must be an anode, a cathode, and an electrolyte. In a corrosion cell, the anode is the negative electrode where corrosion occurs (oxidation), the cathode is the positive electrode where no corrosion occurs (reduction), and the electrolyte is an ionic conductor, usually an aqueous solution.

This tiny corrosion cell is duplicated many times over a steel surface so that the eye sees what appears to be a uniform layer of rust which in reality is a series of multiple corrosion sites. As corrosion products build up over an anode, the surface can at this point become less active electrically and anodes and cathodes can reverse roles. The process will continue until all of the steel is converted to rust, or something happens to break the external circuit.

Coatings in Corrosion Control

This electrochemical rust-forming circuit can be broken by using a barrier. Not only steel may be protected in this manner but other substrates such as concrete and wood may also be shielded from the environment by a barrier. Protective coatings, which serve as barriers, are undoubtedly the engineers' principal materials for easily creating the barrier and giving the desired insulation.

A coating may be defined as a material which is applied to a surface as a fluid and which forms, by chemical and physical processes, a solid continuous film bonded to the surface.

Composition of Coatings

Most coatings consist of three principal parts: pigment, binder, and solvent which dictate the ultimate protection and performance of the coating. The components of a coating all interact to accomplish the purpose for which the coating was designed. The pigments and binder make up the solids of the coating that remain after the coating has dried. The pigments obscure the coated surface, contribute color, prevent premature degradation of the binder by ultraviolet light, and inhibit corrosion. The solvent reduces the viscosity of the coating to permit application—and then evaporates. The binder proceeds from the liquid to the solid state, and in so doing, isolates the surface being coated from the elements.

The interrelationship of coating components is illustrated in Fig. 2-1

Pigments are included in coatings to perform any one or a combination of the following functions:

1. Adding opacity or hiding power
2. Adding color
3. Giving corrosion inhibition
4. Providing resistance to light, heat, water, and chemicals
5. Adjusting the flow properties of the wet coatings
6. Contributing strength

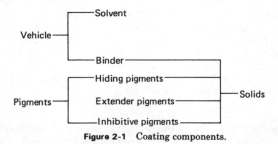

Figure 2-1 Coating components.

Contribution of Each Coating Component

Pigments

Pigments whose principal function is to contribute opacity to the coating are called "hiding" or "prime" pigments. The principal white hiding pigment is titanium dioxide. There are hundreds of colored hiding pigments which, when used singly and in groups, give coatings a variety of colors.

Pigments used to reduce or prevent the corrosion of the coated surface are called inhibitive pigments. The use of inhibitive pigments in modern industrial coatings brings to the war against corrosion all of the principles of corrosion control, including cathodic protection, passivation of the coated surface, and creation of barrier films.

Pigments protect the binder in the coating from the degrading effects of solar radiation. All pigments do this to some degree, but the hiding pigments do it best. Pigments in the dry binder film also reduce the permeability to corrosive elements. Furthermore, pigments add strength and abrasion resistance to coatings. Some pigments are also used to adjust the consistency of coatings so that they may be applied more effectively.

Binders

The binder portion of the vehicle is the component responsible for the curing process as well as for defining basic performance properties. Traditionally these binders have been natural oils derived from plant sources. Some of the more familiar ones are linseed oil and tung oil. Although oil-based coatings are still used in architectural paints, they have been almost totally replaced by synthetic-resin-based coatings for use in heavy-duty applications. These newer coatings dry faster, show better chemical resistance, and give longer service life.

Synthetic-resin binders evolve from the liquid to the solid state by several different drying or cure mechanisms:

1. Solvent evaporation, generally occurring in coatings known as lacquers
2. Drying by a chemical reaction with atmospheric oxygen which is known as oxidation
3. Drying of two-component coatings by a variety of chemical reactions involving each of the components

The simplest drying mechanism is the physical process of evaporation. As solvent evaporates, polymer molecules become more compacted and coil and intertwine. These polymer molecules eventually become bonded together and adhere to the substrate. As an example, vinyl and chlorinated rubber-based coatings dry by this mechanism and demonstrate excellent moisture and chemical resistance. Because this drying or cure process is only physical, not chemical, this type of coating will soften at elevated temperatures and redissolve in strong solvents, the classical definition of a lacquer.

Oil-based coatings, or synthetic binders modified with drying oils, dry by a chemical reaction with atmospheric oxygen. Atmospheric oxygen creates active cross-linking sites of the drying-oil component; these sites connect to form a three-dimensional chemically bonded network. Alkyd and epoxy ester–type binders are examples of systems that cure by oxidation.

The third type of drying or cure mechanism is characteristic of two-component coatings. Two reactive materials, a resin and a polyfunctional cross-linking agent, are mixed prior to use. The functional cross-linking agent reacts with the resin molecules, forming a chemically bonded matrix. Because this type of coating is more chemical- and heat-stable, two-component coatings can be formulated to give heat and solvent resistance, as well as chemical resistance. Two-component epoxy and urethane coatings are representative of this type of drying and cure mechanism.

Solvents

The other portion of the vehicle is the solvent. A solvent dissolves the essential substances in the formulation. It breaks them down and holds the molecular units more or less evenly distributed throughout the solution. The purpose of the solvent is to adjust the flow and viscosity properties of the coating so that it may be successfully brushed, rolled, or sprayed on a surface. In every case, the solvent is designed only to be a carrier for the coating during application and then to evaporate from the coating prior to the completion of the drying cycle.

Volatility or evaporation of solvents is an important property to be considered in the selection of solvents for coatings. The volatility of the solvent determines the drying time of lacquer-type coatings and the tack-time—a time after which flow no longer occurs—of chemically cured coatings. A highly volatile solvent often renders a coating unsprayable by causing the coating to dry before it can touch the surface to be painted—resulting in a phenomenon known as dry spray. A solvent of low volatility often limits the practical applied film thickness of the coating by extending the tack time excessively and causing the coating to sag or run.

SURFACE PREPARATION

Reasons for Surface Preparation

No matter how carefully a coating is formulated and manufactured, or how sound the research on which it was developed, or how sophisticated its chemical technology, the coating will fail prematurely in service if the surface to which it is applied is inadequately prepared. No coating can form a strong bond with a surface if there is something beneath the coating that is weakly bonded to that surface. Dirt, rust, scale, oil, wax, moisture, or other foreign matter provide a weak foundation to hold a paint or coating and consequently cause loss of adhesion. Impurities such as oil or water can prevent adhesion even when present in such small quantities as to be invisible. *Proper surface preparation is vital for best service-life results.*

Types of Substrates

Steel

Probably the most widely used structural material in industrial plants is steel, and it is the surface most frequently encountered in maintenance operations involving paint. Beginning with new steel construction, let us examine some of the surface-preparation problems which are usually encountered when applying a protective paint coating to its surface. Mill scale is one of the most prevalent contaminants found on new, rolled-steel plates, bars, beams, etc.

Mill scale is a hard, blue-black layer formed on steel during the hot-rolling process. It often adheres tightly to new steel, but usually loosens because it is brittle. Mill scale must be removed for best painting results.

Red rust (iron oxide) is another contaminant that is familiar to anyone who has seen an unprotected piece of steel outdoors. It varies in color from bright red to deep brown and may be loose and powdery or hard and brittle. In any case, it also provides a weak foundation for paint, contributes to the breakdown of a coating applied directly over it, and promotes further corrosion if it is painted over rather than removed. For proper coating performance, *all rust* must be removed.

Before proceeding with the surface cleaning of steel, the surface must be inspected for imperfections and troublesome areas, then corrected as follows:

1. Rough welds and other sharp edges must be ground smooth.
2. Weld spatter must be knocked off with a scraper, chisel, or other means.
3. Seams joined with skip or tack welds must be rewelded with a continuous, smooth bead.
4. Rivets and bolts must be set firm and tight, or caulked if they are angled or not flush.
5. Crevices and pits should be caulked.

Welding should be done before the cleaning operation, and caulking after the cleaning operation.

Choose a surface preparation method that will clean the steel and be within the limits of cost, accessibility, contamination of the manufacturing or cleaning process, damage to machinery and equipment, and physical harm to personnel.

Abrasive Blast Cleaning. Blast cleaning is the best method for cleaning steel because it is effective in removing all scale and corrosion products, as well as chemical contaminants. It also provides an anchor pattern in the steel for the coating due to the roughness of the blast pattern and the increase in surface area. The various types of blasting abrasives which are used perform differently, and produce characteristically different types of surface pattern. The pattern, or profile, resulting from blasting is extremely important in its effect on a coating's performance.

As an example, if the blasted surface is too smooth, there will not be adequate "anchor" for the coating. On the other hand, if the surface is too rough, sharp pinnacles of metal are likely to project through the coating and the steel will remain unprotected. Sand is the most widely used abrasive because of its availability and low cost. It is important to use hard sand that does not produce excessive dust. It is also important to use a clean sand free of clay and other foreign matter. In other words, a washed, graded silica sand should be used.

Conventional sandblasting procedure is to use dry sand and dry air. However, "wet" blasting is sometimes used because dusting is thereby reduced. In wet blasting, water is injected at the nozzle or is mixed with sand in the pot. In this process, very little dust is produced, but a surface covered with wet sand remains, and this must be washed off. Inhibitors are used in the water to prevent the steel from flash rusting.

Steel grit is sometimes used in place of sand. These grit particles are hard, angular-shaped pieces of cast iron, malleable iron, or steel with sharp edges. Steel grit has several advantages over sand; it produces less dust, cuts faster, and can be reclaimed for future use. Its disadvantages are: (1) higher cost and (2) a tendency to roughen the metal excessively. Again, like sand, the proper size of steel grit must be selected. Too fine a grit may not give a good pattern; conversely, coarse grit will cut deeply into the surface, leaving points of metal sticking up. In addition, steel grit used over an extended period under humid conditions will eventually rust—contaminating rather than cleaning the surface.

Steel or iron shot may also be used as an abrasive. But shot blasting is relatively inefficient because the particles are round and smooth and tend to hammer foreign matter into the surface without removing it.

Table 2-1 shows the maximum height of profile produced by various types of abrasives in blast-cleaning steel.

Surface Preparation Procedures. There are accepted standards that cover the cleaning of steel by several methods including abrasive blasting. These standards have been established by the Steel Structures Painting Council (SSPC) and the National Association of Corrosion Engineers (NACE). The following paragraphs are summaries of these standards.

TABLE 2-1   Characteristics of Various Blasting Media

Abrasive	Maximum particle size	Maximum height of profile	
		mils	μm
Sand, very fine	Through 80 mesh*	1.5	38.1
Sand, fine	Through 40 mesh	1.9	48.3
Sand, medium	Through 18 mesh	2.5	63.5
Sand, large	Through 12 mesh	2.8	71.1
Crushed iron grit #G-50	Through 25 mesh	3.3	83.8
Crushed iron grit #G-40	Through 18 mesh	3.6	91.4
Crushed iron grit #G-25	Through 16 mesh	4.0	101.6
Crushed iron grit #G-16	Through 12 mesh	8.0	203.2
Iron shot #S-230	Through 18 mesh	3.0	76.2
Iron shot #S-330	Through 16 mesh	3.3	83.8
Iron shot #S-390	Through 14 mesh	3.6	91.4

*U.S. Sieve series.

Solvent Cleaning: SSPC-SP 1-63. Solvents, emulsions, cleaning compounds, steam cleaning, or similar materials and methods are used to remove oil, grease, dirt, drawing compounds, and other similar foreign matter from surfaces prior to painting. Certain corrosive salts, such as chlorides and sulfates, must be removed with water before cleaning the surface with hydrocarbon-type solvents.

When rags or waste are used with solvents for cleaning, they should be replaced frequently with clean ones, since they absorb grease and then act as a transportation agent for the grease or other contaminants. Use plenty of clean rags, and as an added safety measure, wipe the surface with clean saturated rags.

Solvent cleaning should be undertaken only with adequate ventilation. Caution should always be observed, because of the possibility of fire.

Hand-Tool Cleaning: SSPC-SP 2-63. Hand-tool cleaning is generally confined to wire-brushing, scraping, chipping, and sanding. Surface preparation of this quality is not recommended if finishing systems with poor wetting characteristics such as vinyl coatings are to be applied. This type of cleaning is normally used to prepare the surface in areas where corrosion is a minor problem. On weld seams or spots, always remove the surrounding weld flux spatter, since it promotes premature paint failure.

Power-Tool Cleaning: SSPC-SP 3-63. Power tools of various kinds and shapes, either electric or pneumatic, are used for this type of surface preparation. Rotary wire brushes, chipping hammers, abrasive grinders, descalers, needle guns, and sanding machines are included in the equipment required for power-tool surface preparation.

This method of surface preparation is often used to remove rust and scale left on the surface after other methods have been used to remove the bulk of the contamination. Care must always be exercised; inexperienced workers will sometimes mistake burnished mill scale for bright clean metal. The primer will not adhere to burnished areas of this kind, and premature paint film failure results.

Power-tool cleaning is considered to be somewhat more efficient than hand cleaning if care and good judgment are used by the mechanics.

Flame Cleaning of New Steel: SSPC-SP 4-63. One traverse of the flame-cleaning head, followed by wire-brushing, is normally sufficient to remove rust and scale from new or old steel that has been partially weathered. Badly rusted or heavily scaled steel should be treated as previously painted or "old work." At best, flame cleaning will give results somewhere between that obtained with power tools and that from commercial blast cleaning. The individual job determines the types of gas and the speed of traverse required for acceptable cleaning.

White-Metal Blast Cleaning: SSPC-SP 5-63 (NACE No. 1). This system provides the maximum surface preparation and should result in the best performance possible from the painting system chosen. Surface preparation of this magnitude is often done in the field after the equipment or structure is in place.

This system of cleaning is mandatory for preparing the interior of tanks prior to the application of a lining. The cost is comparatively high, and it is usually used for work only when such costs are warranted. In some cases, accessibility after erection is not possible; therefore, under such conditions, the use of blasting to white metal before erection is considered economical.

The use of white-metal blast cleaning in areas of high humidity without having the steel rust back necessitates the choice of proper blasting conditions so that no rust-back can occur—and painting must be completed, at least, on the same day. A white-metal blast-cleaned surface finish is defined by the SSPC as a surface with a gray-white, uniform metallic color, slightly roughened to form a suitable anchor pattern for coatings. When viewed without magnification, the surface should be free of all oil, grease, dirt, visible mill scale, rust, corrosion products, oxides, paint, or any other foreign matter.

Commercial Blast Cleaning: SSPC-SP 6-63 (NACE No. 3). This type of blast cleaning is generally considered adequate for most surfaces requiring a normally clean surface for painting. It is also economical and generally recommended for surfaces requiring fast-drying protective coatings.

A commercial blast-cleaned surface finish is defined as one in which all oil, grease, dirt, rust scale, and foreign matter have been completely removed from the surface, and all rust, mill scale, and old paint have been completely removed except for slight shadows, streaks, or discolorations caused by rust stain, mill scale, oxides, or slight, tight residues of paint or coating that may remain. If the surface is pitted, slight residues, rust, or paint may be found in the bottom of pits. At least two-thirds of each square inch of surface area should be free of all visible residues, and the remainder should be limited to the light discoloration, slight staining, or light residues mentioned above.

Brush-Off Blast Cleaning: SSPC-SP 7-63 (NACE No. 4). This is a relatively low-cost method of cleaning and is often used at the jobsite to clean materials that are shop- and field-coated before installation. This type of cleaning is not usually recommended for severe environments. It is often used to remove temporary coatings applied for the protection of equipment in transit or storage. It is also used to remove old finishes in bad condition.

A brush-off blast-cleaned surface finish is defined by the SSPC as one from which all oil, grease, dirt, rust-scale, loose mill scale, loose rust, and loose paint or coatings are removed completely. However, tight mill scale and tightly adhering rust, paint, and coatings are permitted to remain—provided that all mill scale and rust have been exposed to the abrasive blast pattern sufficiently to expose numerous flecks of the underlying metal fairly uniformly distributed over the entire surface.

Pickling: SSPC-SP 8-63. Pickling is a method of preparing metal surfaces for painting by completely removing all mill scale, rust, and rust scale by chemical reaction, electrolysis, or both. It is intended that the pickled surface should be completely free of all scale, rust, and foreign matter. Furthermore, the surface should be free of unreacted or harmful acid, alkali, or smut.

Near-White Blast Cleaning: SSPC-SP 10-63T (NACE No. 2). Near-white blast cleaning is a method of preparing metal surfaces for painting or coating by removing nearly all mill scale, rust, rust scale, paint, or foreign matter by the use of abrasives propelled through nozzles or by centrifugal wheels.

A near-white blast-cleaned surface finish is defined as one in which all oil, grease, dirt, mill scale, rust corrosion products, oxides, paint, or other foreign matter have been completely removed from the surface except for very light shadows, very slight streaks, or slight discolorations caused by rust stain, mill scale oxides, or slight, tight residues of paint or coating that may remain. At least 95 percent of each square inch of surface area should be free of all visible residues, and the remainder should be limited to the light discolorations mentioned above.

The complete specification for these procedures may be found in the Steel Structures

Painting Council publication, *Good Painting Practice.*[1] Pictorial standards for these procedures, SSPC-VIS-1, are also available from the Steel Structures Painting Council (see "Specifications and Standards.")

Other Substrates

In addition to steel, there are other substrates that may be encountered occasionally in maintenance operations. These surfaces also must be prepared properly to receive a coating system.

Cast Iron.   Cast iron is a porous material and is likely to absorb moisture or other liquids with which it comes in contact. To drive absorbed material out of its pores, cast iron should be heated before sand or grit blasting. This can be done by placing it in an oven for 8 to 12 h at 300°F (149°C) or by heating it with torches until this temperature is reached.

Zinc.   New zinc surfaces, galvanized or metal-sprayed, should first be wiped with solvent. The clean surface is then treated with a specially formulated acid-treatment solution. Such solutions are designed to etch zinc, thereby providing sufficient "tooth" for the application of coatings. However, if zinc-coated surfaces are allowed to properly weather, sufficient roughness is obtained to permit direct coating without prior acid treating or sand blasting.

Copper or Brass.   These metals should be blasted lightly to remove oxides and provide "tooth" for the coating.

Aluminum.   Aluminum surfaces must first be degreased by solvent cleaning. The clean surface may then be cleaned with either a chromate-type conversion treatment, a phosphate chemical treatment, or a wash primer treatment.

Concrete.   Concrete is coated to protect it from chemical attack or physical damage by spalling and cracking. There are several factors which must be considered when preparing concrete to receive a coating system:

1. Laitance is a thin layer of incompletely hydrated cement which floats to the surface. Since this layer has poor strength and adheres loosely to the concrete, it must be removed. This may be accomplished by either sand blasting or acid etching.
2. Efflorescence is a deposition of salts on the concrete surface, caused by moisture released during the curing process. These deposits are alkaline and must be removed by acid etching.
3. Form oils, used for easy removal of concrete pouring forms, present problems because a coating will not wet the oils. These oils must be removed by detergent cleaning before acid etching or sand blasting the surface.
4. Concrete hardeners are sometimes used to increase surface hardness and decrease the permeability of concrete. They migrate to the surface and cannot be removed by acid etching. They must be removed by sand blasting.

How to Etch Concrete.   Carefully prepare a solution consisting of one part full-strength muriatic acid, mixed with three parts water. Use 1 gal of solution to 100 ft² (9.29 m²) of floor, and scrub well while applying. Be sure to use enamel, plastic, or wooden pails. The acid *must be added* to the water in order to prevent adverse effects. Allow the solution to remain on the floor until it stops bubbling. Scrub the surface well, using a stiff bristle brush. Flush off thoroughly with clean water. After the surface has dried (less than 15 percent moisture), it is ready for painting. *Use appropriate eye- and skin-protective devices.*

USE OF PROTECTIVE COATINGS

Systems Concept

There is a right way and a wrong way to select coatings. For instance, a coat of paint can be "slopped" on every couple of years, or a system can be engineered that will last for a much longer time.

Coatings are very selective with respect to the types of environment and corrosion they are capable of protecting against. They are equally selective concerning some of their other properties such as impact resistance, abrasion resistance, flexibility, bondability to a substrate, and their appearance. *No one generic class of coatings is universally acceptable.* Sometimes there can be a dilemma—such as the one in which the coating provides the protection required for use on catwalks but does not give adequate abrasion resistance.

In most cases, the best result can be obtained by combining two, and sometimes more, coatings into a coating system. The typical coating system usually consists of a primer applied to the metal surface and a top coat. Primers are selected for these characteristics:

1. Bond to metal surface
2. Rust-inhibitive pigment content

Topcoats are usually characterized by:

1. Attractive appearance
2. Color retention and resistance to uv radiation from the sun
3. Low permeability to moisture, chemicals, etc.
4. Abrasion and impact resistance
5. Chemical resistance

With the variety of coatings available today, and the many painting systems possible with these coatings, the concept of the coating system adds versatility to paint technology. With the greater selection available in the coating-system concept, there is now an excellent chance that the user can zero in on the exact protection required without paying a premium for overprotection. See Ref. 2.

In order to arrive at the best coating system for the purpose, certain information must be available, including:

1. Type of surface to be coated
2. Type of surface preparation that can be achieved
3. Type of chemical exposure (if any) expected
4. Operating temperature of surface to be coated
5. Type of structure being coated
6. Degree of abrasion and impact resistance desired

Coating Characteristics

It would be difficult to thoroughly discuss all of the characteristics of the many coating types which are available. Table 2-2 summarizes the important properties of some of the more common coating types. These properties would be considered the optimum to be expected from the coatings. Should more detailed or specific coating information be required, it would be wise to contact a reputable coatings supplier.

TABLE 2-2 Coating Characteristics

Coating type	Type of drying mechanism	Resistance*					Max. drying temperature		Remarks
		Mineral acid	Alkali	Solvent	Water	Weather	°F	°C	
Epoxy	Catalyzed (two-component)	Good	Excellent	Fair	Good	Good (chalks)	300	149	
Vinyl	Solvent evaporation	Excellent	Good	Poor	Very good	Very good	120	49	
Chlorinated rubber	Solvent evaporation	Good	Good	Poor	Good	Good	120	49	
Urethane	Catalyzed (two-component)	Excellent	Excellent	Fair	Good	Excellent	300	149	
Silicone	Catalyzed (heat)	Very good	Fair	Poor	Excellent	Excellent	1000	538	
Alkyd	Oxidation	Poor	Poor	Poor	Poor	Good	180	82	
Silicone alkyd	Oxidation	Poor	Poor	Poor	Fair	Very good	300	149	Min 25% silicone
Epoxy ester	Oxidation	Fair	Fair	Poor	Fair	Good (chalks)	250	121	Min 45% epoxy
Coal tar–epoxy	Catalyzed (two-component)	Very good	Excellent	Fair (bleeds)	Excellent	Fair (chalks and bronzes)	325	163	
Inorganic zinc	Hydration	Poor	Poor	Excellent	Excellent	Good (6–8 pH)	750	399	
Bitumastic	Solvent evaporation	Good	Good	Poor	Very good	Poor	150	65	Weatherable grades available

*Splash and spillage

Economics

Cost of Material

Up to this point, the systems concept or coating systems engineered for corrosion control has been discussed; but the economics of painting must also be considered. All too often persons purchasing coatings emphasize the cost per gallon. While the cost per gallon or per liter is a consideration, the decision could change when the costs are based on the volume solids and spreading rate. The following example will illustrate this.

	Cost, $/Gal	Volume solids, %	Dry-film thickness	
			mils	μm
Coating A	4.50	30	4	101.6
Coating B	4.10	25	4	101.6

Coating A

$$\frac{1604 \times 0.3 \text{ (volume solids)}}{4 \text{ mils}} = 122 \text{ ft}^2/\text{gal } (2.9 \text{ m}^2/\text{L})$$

where 1604 is a constant and

$$\frac{\$4.50/\text{gal}}{122 \text{ ft}^2/\text{gal}} = \$0.037/\text{ft}^2 \ (\$0.398/\text{m}^2)$$

Coating B

$$\frac{1604 \times 0.25 \text{ (volume solids)}}{4 \text{ mils}} = 101 \text{ ft}^2/\text{gal } (2.5 \text{ m}^2/\text{L})$$

where 1604 = a constant

$$\frac{\$4.10/\text{gal}}{101 \text{ ft}^2/\text{gal}} = \$0.04/\text{ft}^2 \ (\$0.431/\text{m}^2)$$

We can see from this example that coating A is the best bargain even though it has a greater cost per gallon. The nomograph shown in Fig. 2-2 will allow you to determine material costs and make comparisons quickly and simply.

System Life

Although these economics are impressive, there is still a tendency to look at "low-cost" systems because of modest painting budgets. A look at the cost breakdown of a typical coating job will show the materials cost.

Application labor	30 to 50%
Surface preparation labor	15 to 40%
Coating material	15 to 20%
Clean-up labor	5 to 10%
Tools and equipment	2 to 5%

In this frame of reference, the difference between a high-cost system and a low-cost system is not really significant. With the cost that goes into labor, regardless of materials costs, it would seem that the high-cost system should be the choice. Figure 2-3 further illustrates this.

Each step in the chart lines is either a recoat or touch-up job. The high-cost or properly engineered system does cost more initially, but the longer periods between touch-up

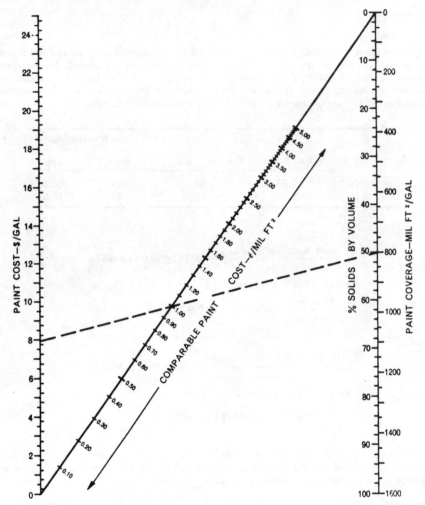

Figure 2-2 Nomograph for calculating paint material costs.

or recoat, as compared to the low-cost system, result in a lower cost of maintaining a structure over a period of years. Quality truly pays.

Inspection

To obtain the planned economies and to realize the maximum potential of a coating system, it is essential that periodic inspections be made before, during, and following the application. The inspection begins with the writing of the specification. Suitable standards for surface preparation, application, and inspection must be established before the job begins. A good specification alone will not guarantee a good coating system. It is necessary for a qualified inspector to see that all specifications are followed and that all defects are promptly remedied. The specification should include the duties and authority of the inspector along with all the quality control or measuring equipment to be used.

A successful coating system must have:

1. Proper film thickness for long-term durability
2. Coating continuity (freedom from holes)

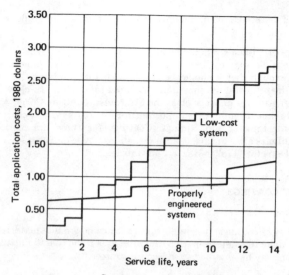

Figure 2-3   Service life vs. application costs.

3. Good adherence, or bonding, to the substrate
4. The ability to cure under proper conditions

Surface Preparation. Before the cleaning operation begins, all parties concerned with a particular job should have a clear understanding of the specified surface preparation. It is recommended that Steel Structures Painting Council specifications (see "Specifications") be used. These specifications may be found in SSPC's publication *Good Painting Practice*.[1]

All cleaning equipment should be inspected before the job begins. The surface cleaning operation should be inspected periodically to ensure compliance with health, safety, and quality of work. To aid in this inspection, there are tools available such as:

1. Steel Structures Painting Council Visual Standards, SSPS-VIS-1
2. CAPS (Clemtex Anchor Pattern Standards)

All cleaned surfaces should be coated before the end of the workday.

Application. Inspection of the application process actually begins with the materials. Records should be made of the code number and batch number of the coatings and of the area where they were applied. Material should be inspected for such deficiencies as skinning, thickening, gassing, gelling, and excessive settling. The painting materials should be mixed and thinned according to manufacturer's directions using mechanical mixers. Special attention should be given to two-component materials to ensure proper mixing, and the painter should have a thorough knowledge of the products' digestion time and pot life. In addition, application equipment should be inspected to ensure that it is in good working order and meets the requirements for the coating being applied.

The application of all coatings should be followed closely by an inspector making sure the equipment and material are behaving properly and the correct amount is applied. Particular attention should be given to adequate coverage on sharp edges, rivets, corners, and crevices. The amount of material applied can be checked with an instrument such as a Nordson prong-type gauge which reads wet-film thickness.

Weather conditions form another important factor in the paint application process. In general, coatings should not be applied at metal temperatures below 40°F (4°C) nor above 125°F (52°C) and at a temperature at least 5°F (2°C) above the dew point. In the case of catalyzed materials, the minimum application temperature is 60°F (15°C) for the metal being coated.

Dry Film.   The dry film also should be inspected for defects such as peeling, blistering, pinholing, fish-eyeing, sagging, blushing, and failure to dry. In some cases, the defect may be serious enough that it should be corrected before applying the next coat. Each coat, as well as the total system, should be measured for dry-film thickness. This may be done with a magnetic type gauge such as the Mikrotest (Nordson Corp., Amherst, Ohio).

While the absence of pinholes is important for all coatings, it is critical for coatings being applied for tank or pipe linings. Coatings for these applications should be inspected for pinholes after drying, using a wet-sponge-type detector such as the Tinker-Razor Holiday Detector (Tinker-Razor, San Gabriel, Calif.).

APPLICATION OF COATINGS

Estimating Coverage

In the application of coatings, it is important to know how much material is needed for a given job. To make this determination, the total area to be coated must first be known. This information may be obtained from several sources.

A. Estimating from Weight of Steel:

  Steel Plate.   98 ft² (3.6 m²) per inch (centimeter) of thickness, both sides; for example, ½-in-(1.3-cm-) thick steel plate = 196 ft²/ton (20.1 m²/t), both sides

  Angles. $3 \times 3 \times$ ½ in @ 9.4 lb/ft = 213 ft²/ton ($7.6 \times 7.6 \times 1.3$ cm @ 14.1 kg/m = 21.8 m²/t)

  Channels.   6 in @ 8.2 lb/ft = 390 ft²/ton (15.2 cm. @ 12.2 kg/m = 39.9 m²/t)

  I Beams.   12 in @ 35 lb/ft = 211 ft²/ton (30.5 cm @ 52.1 kg/m = 21.6 m²/t)

  Piping.   4 in Std. @ 10.8 lb/ft = 219 ft²/ton (10.2 cm Std. @ 16.1 kg/m = 22.4 m²/t)

  These are examples of the type if information found in steel company manuals or in the AISC (American Institute of Steel Construction) manual.

B. Estimating by Weight and Type of Structure

  Structural steel, after erection, generally falls into classifications which may be readily identified:

  1. Light structural steel, for example, would be represented by electric transmission towers and installations of that type.
  2. Heavy structural steel is the equivalent of built-up bridge girders or designs which utilize built-up and fabricated heavy structural members.
  3. Extra-heavy structural steel is used in structures which utilize large rolled structural shapes.
  4. Medium or mixed structural steel will be the combination normally seen in trusses, etc., where a mixture of heavy to very heavy compression members are combined with substantial amounts of much lighter cross bracing.

  Using these descriptions, the average surface area per ton of steel for different types of construction is shown in Table 2-3.

C. Geometric and Equipment-Surface Relationships

  Certain geometric relationships are repeated for convenience.

  1. The area of a triangle is its base times one-half its altitude ($a = $ ½ bh).
  2. The area of a square or a rectangle is the product of two adjacent sides ($a = hw$).
  3. The inside or outside of a cylinder (straight sides) is the product of its height, its diameter, and π (3.1416) ($a = h\pi d$).

TABLE 2-3   Average Surface Area of Steel

Type of construction	Area, ft² per ton	Average wt, U.S. tons	Area, mt² per ton	Average wt, metric tons
Light	300–500	400	30.7–51.2	41.0
Medium	150–300	225	15.4–30.7	23.0
Heavy	100–150	125	10.2–15.4	12.8
Extra heavy	50–100	75	5.1–10.2	7.7
Average industrial plants	200–250	225	20.5–25.6	23.0

4. The area of a dished head, such as a pressure vessel end, is usually 1.58 times the area of a flat circle ($a = 1.58\pi r^2$).

In addition, certain other relationships are helpful:

1. The area of a standard corrugated sheet with ribs 2½ or 1¼ in (6.4 or 3.2 cm) in size is 8 or 11 percent greater, respectively, than the area it covers.
2. Open wet steel joists should be considered as solid rather than open; double the measurement to account for both sides and add 10 percent.
3. In structural steel, add 20 percent for rivets, plates, and flanges. A 5 percent allowance for plates and flanges is adequate for welded structures.
4. Gratings may be approximated by multiplying the area covered by 4 to 6 (varying according to depth). Use an average of 5 for most gratings.
5. For window sash, allow 3 ft² (0.3 m²) per pane.
6. For large valves, allow 20 ft² (1.8 m²) per valve [above 4 in (10.2 cm)].

Another factor that must be considered in estimating coverage is the amount of material loss during the application process. The following are loss ranges for different methods of application:

Application	Loss, %
Brush	4–8
Roller	4–8
Conventional spray	20–40
Airless spray	10–20

The amount of loss will vary with the size and shape of the surface and the environmental conditions. Under adverse conditions of wind and small surfaces, spray loss can be 50 percent or more.

Now that we have several means of estimating area to be coated, we must determine the amount of paint needed. The following example will illustrate the steps to accomplish this:

Calculate the amount of paint needed to apply the finish coat, at a thickness of 2 mils dry, on the exterior of storage tank. The storage tank is constructed with 50 tons (45.5 t) of ½-in (1.3-cm) steel plate. The finish coat is an epoxy with 50 percent volume solids and is applied by airless spray.

Exterior surface of tank:

½-in (1.3-cm) steel plate, 98 ft²/ton (10 m²/t) per side

50 tons (45.5 t) $\times$ 98 ft²/ton (10 m²/t) = 4900 ft² (455 m²)

Coverage rate of paint:

$$\frac{1604 \times \text{volume solids}}{\text{dry film thickness}} = \text{theoretical coverage}$$

$$\frac{1604 \times 0.5}{2} = \frac{802}{2} = 401 \text{ ft}^2/\text{gal} \ (9.8 \text{ m}^2/\text{L})$$

Allowance for spray loss (20 percent):

$401 \text{ ft}^2/\text{gal} \ (9.8 \text{ m}^2/\text{L}) \times 80\% = 320.8 \text{ ft}^2/\text{gal} \ (7.8 \text{ m}^2/\text{L})$

Amount of paint needed:

$$\frac{4900 \text{ ft}^2 (455 \text{ m}^2)}{320.8 \text{ ft}^2/\text{gal} \ (7.8 \text{ m}^2/\text{L})} = 15.27 \text{ gal} \ (58.3 \text{ L})$$

Application

Application of coating materials not only involves application methods and equipment but also includes other factors of environmental conditions, materials preparation, film characteristics, curing, and recoating. These factors are generally covered in the coating manufacturers' literature. This literature should be read and understood before the job begins. Do not deviate from these instructions or take short cuts. This can only lead to problems that will be difficult and costly to resolve.

Application Equipment

Conventional Air Sprays. In conventional air spraying, paint is pushed through a nozzle in a spray gun by air pressure. At the nozzle, the paint is atomized by jets of air. The volume output of paint by air spray is relatively small. A conventional air spray setup consists of a pressure pot, air and fluid hose, spray gun, and air supply. The pressure pot should be equipped with a double regulator, to regulate both fluid pressure and atomizing air, and an agitator. All parts of the spray gun should be tightly connected, as should all hose connections. Any air leak will result in inefficient operation.

Airless Sprays. In airless spraying, paint is pumped mechanically under high pressure through a spray-gun nozzle. The paint is pumped at such a high velocity and pressure that it atomizes as it emerges from the small orifice in the nozzle. The volume of paint moved by airless spray is much greater than by air spray. Airless spray permits the application of higher viscosity materials, which will allow greater film build per coat. An airless spray setup consists of a hydraulic pump, high-pressure fluid hose, spray gun, and air supply.

Brushes and Rollers. Although heavy-duty maintenance coatings are normally applied by spray because it is the best and most efficient method, there are situations when it is impossible to spray-apply coating materials, e.g., in difficult-to-reach areas or areas where spray would be hazardous or damaging to adjacent areas. Brush and roller application is slower, less efficient, and more costly.

Brushes and roller equipment must be selected according to the type of material being applied. When strong solvent materials like epoxies are used, brushes and roller materials should be selected for their resistance to strong solvents.

SPECIFICATIONS AND STANDARDS

Surface Preparation

1. Surface preparation specifications (written).
2. SSPC-VIS, "1 Pictorial Surface Preparation Standards," Steel Structures Painting Council, 4400 Fifth Avenue, Pittsburgh, PA 15213.
3. CAPS—Clemtex Anchor Pattern Standards.
4. "Visual Comparator," Clemtex, Ltd., P.O. Box 15214, Houston, TX 77020.

Performance

1. SSPC-VIS 2, "Standard Methods of Evaluating Degree of Rusting on Painted Steel Surfaces," Steel Structures Painting Council, 4400 Fifth Avenue, Pittsburgh, PA 15213.
2. ASTM D 714, "Blister Standards," American Society for Testing and Materials, 1916 Race St., Philadelphia, PA 19103.

REFERENCES AND BIBLIOGRAPHY

1. *Steel Structures Painting Manual,* vol. 1, *Good Painting Practice,* vol. 2, *System and Specifications,* Steel Structures Painting Council, 4400 Fifth Avenue, Pittsburgh, PA 15213, 1966.
2. *Industrial Maintenance Painting,* National Association of Corrosion Engineers, P.O. Box 986, Katy, TX 77450, 1973.

NACE Standards and Technical Committee Reports, National Association of Corrosion Engineers, P.O. Box 986, Katy, TX 77450, in the following subject areas:

Surface Preparation for Protective Coatings

Protective Coatings for Atmospheric Service

Coatings and Linings for Immersion Service

Specialty Coatings

Painting and Decorating Craftsman's Manual and Textbook, Painting and Decorating Contractors of America, 7223 Lee Highway, Falls Church, VA 22046, 1978.

Uhlig H. H. (ed.): *The Corrosion Handbook,* Wiley, New York, 1948.

NACE Basic Corrosion Course Text, National Association of Corrosion Engineers, P.O. Box 986, Katy, TX 77450, 1977.

Index

ABOUT THE EDITORS

Editor in Chief ROBERT C. ROSALER, P.E., is Vice President of James O. Rice Associates, Inc., organizers of practical engineering conferences and seminars. He is a member of the American Institute of Plant Engineers and the Plant Engineering Division of the American Society of Mechanical Engineers. He has over 40 years of experience as an engineer executive with both large and small corporations.

Associate Editor JAMES O. RICE has been identified with the professional management movement since 1935. A former Vice President and General Manager of the American Management Association, where he introduced the seminar system of education, Mr. Rice is the founder and President of James O. Rice Associates, Inc.